TORAH &&
TECH

Discussions at the Intersection of Torah and Technology

RABBI BEN GREENBERG

RABBI YECHIEL KALMENSON

Published by Torah && Tech Publishers

Copyright © 2020 by Ben Greenberg and Yechiel Kalmenson

Published in the United States by IngramSpark.

www.torahandtech.dev

ISBNs
978-1-7350217-0-6
978-1-7350217-1-3

Manufactured in the United States of America

First Edition: June 2020

TABLE OF CONTENTS

1. About This Book...vii

2. Acknowledgments...ix

3. Introduction..xi

Genesis - Bereshit

1. Bereshit Object-Oriented Genesis..2

2. Noach The Floods of Email...5

3. Lech Lecha Are You a 10x Engineer?...7

4. Vayera The Headphones Can Come Off Sometimes..................9

5. Chayei Sara On Artificial (Un)intelligence..............................11

6. Vayetzei Maintaining the Proper Headspace............................16

7. Vayishlach All In The Details..18

8. Vayeshev Family: It's Complicated!...20

9. Vayigash Virtual Identities...22

10. Vayechi The Case for Zevulun...25

Exodus - Shemot

1. Shemot Whose Responsibility Is It? 30

2. Bo We're Free... Let's Make a Calendar!.......................32

3. Beshalach Keep Calm… and Go for a Walk....................34

4. Yitro You Cannot Do It Alone 37

5. Mishpatim Am I My Code's Keeper? 39

6. Terumah The Mentorship Relationship 41

7. Tetzaveh What's the Point of Pointers? 43

8. Ki Tisa What Is Your Technical Debt 46

9. Vayakhel Don't Lose Focus! 48

10. Pekudei Continuous Integration in the Wilderness...............50

Leviticus - Vayikra

1. Vayikra When Smaller Is Bigger 54

2. Tzav Encountering Burnout 56

3. Shemini To Err Is Divine...59

4. Tazria Dealing With Unhealthy Environments....................62

5. Metzora Providing Empathetic Feedback.......................64

6. Acharei Mot We Can Do Better! 67

7. Kedoshim Finding Unity With Our Differences 70

8. Emor I Mentor Therefore I Am..73

9. Behar A Little Help Can Go a Long Way! 75

Numbers - Bamidbar

1. Bamidbar A More Compassionate Workplace.........................80

2. Nasso Value Every Contribution 82

3. Behaalotecha Do Your Part...84

4. Shlach Challah Is More Than Bread..86

5. Korach Holy Tech!...88

6. Chukat Brute Force A Solution?..90

7. Balak Bilaam the Hacker 92

8. Pinchas Advocate for Yourself..95

9. Matot-Masei See Something? Say Something!........................97

Deuteronomy - Devarim

1. Devarim Yes, You Can... But, First It's Okay to Cry 100

2. Va'etchanan Bounce Back! 102

3. Eikev You Are More Than Your Job 104

4. Re'eh Elul - Our Annual Retro... 106

5. Ki Teitzei Am I My Code's Keeper? 108

6. Ki Tavo Think (Also) With Your Heart.................................... 110

7. Nitzavim Feedback? It's All in How You Express It.............................. 112

8. Haazinu You Belong! 114

Holidays

1. Passover Matzah and Keeping It Simple................................. 118

2. Shavuot A Different Perspective 120

3. Sukkot Embrace Fragility 123

4. Epilogue: One Year and Beyond... 125

5. About The Authors 127

1

ABOUT THIS BOOK

This book is the culmination of one year of a project that we began in earnest prior to Chanukah of 2019. The goal of the project was simple, while the topic was, and remains, complex. The goal was to produce a weekly newsletter that brought Torah -- Jewish wisdom and ethics -- to bear on tech, and tech to bear on Torah.

Jewish tradition has been in a dialogue with the world around it since the first words of Torah were hearkened at Mt. Sinai some 3,500 years ago. The newsletter aimed to bring a small, but hopefully meaningful, contribution to that dialogue in the space between Torah and technology.

We named it Torah && Tech precisely because we believe that the proposition of an ethics and wisdom tradition pairing with technology is not an either/or proposition. Instead, it is a clear and unequivocal yes and, represented in programming language syntax by the double ampersand, &&.

The topics addressed in the first year are wide and broad. Each week offers a short and easy to digest nugget on the topic centered mainly around the weekly Torah portion read in synagogue on Shabbat. We know that each short essay on themes as evocative as artificial intelligence, mentorship, responsibility, burnout, feedback, and more cannot possibly reach the depths on those subjects. Yet, we hope that they spark conversations that are sorely needed in the tech community.

We sought to preserve the feel of the newsletter in this book. As such, editing has been kept to a minimum, mainly to clean up grammar and spelling fixes, while the content remains the same.

An astute reader will notice that not every weekly Torah portion is covered in the book. This, too, preserves the original feel of the newsletter. Some weeks were holidays, and they are included in the Holidays section, while other weeks, as a volunteer effort driven by love, we did not manage to publish given other pressing commitments on our time and schedule. We did not think it would be in the interest of the book to "backfill" those chapters and pretend otherwise.

We hope this book serves to ignite conversations for you on the themes laid out in the following pages. Technology has quickly come to not just act as a tool for productivity, but as the primary vehicle for society and culture. We write these words during the height of the coronavirus pandemic, and as the world turns towards technology for everything from school meetings to after-work hangouts, this has never been more true.

We welcome your comments and feedback, and if you find it meaningful, we are still publishing new editions of *Torah && Tech* each week. You can subscribe at www.torahandtech.dev.

-- Yechiel and Ben

2

ACKNOWLEDGMENTS

This book, and the year of Torah && Tech newsletters that preceded it, would not have been possible without so many people and other initiatives that supported and enabled the work.

We want to thank our spouses, Sharon and Shayna, respectively, for providing invaluable feedback on the ideas each week and allowing us the time to dedicate to writing Torah && Tech throughout the year.

We would also like to acknowledge our wonderful guest contributor, Zoe Lang, who brought her unique perspective to the initiative. You can find her words in chapter 1 of the section on Genesis.

This book would not have been possible without phenomenal online tools such as Chabad.org and Sefaria that make Jewish text learning accessible on any device and at any time. We actively utilized their resources for textual analysis.

Last, but certainly not least, we are grateful to Leon Adato for his mentorship in the book publishing process and Sara McCombs for her exceptional work in editing the manuscript.

3

INTRODUCTION

Torah *&& Tech* launched Chanukah 2019, and what follows is the inaugural welcome message.

Welcome to Torah && Tech!

We are very excited to be launching this newsletter, uniting the world of Torah and the world of tech.

It's no coincidence that the launch coincided with Chanukah, the holiday where we celebrate a victory over a way of life that devalued the search for meaning in the day-to-day.

The Seleucid Hellenistic position was epitomized by their oppressive edict, as recorded in the Jewish tradition, "כתבו לכם על קרן השור שאין לכם חלק באלוקי ישראל -- Carve out on the horns of your oxen that you have no part in the God of Israel." This decree seems strange. Why the insistence that the Jewish people carve a renouncement of God on the horns of their oxen? Wouldn't it make more sense to engrave it in their synagogues and yeshivot, houses of study? The places where we go to seek God in a formal and public way?

However, the truth is, the Seleucid Hellenists didn't have a problem with the God of the Jews necessarily, and not even with the Torah, that presented a counterpoint to their overall outlook. The Hellenistic tradition had their own multiplicity of gods. Why would the notion of a God of the Jews bother them?

Indeed, they took no issue with the Torah and with Jews engaging with it, as long as it was confined to the walls of the study halls and synagogues. They bitterly opposed the idea that the Torah could be relevant outside of these confined spaces, that it had a place outside of the parochial.

A central aspect of the cultural zeitgeist then was that philosophy was merely a mental exercise, something to keep the mind occupied with, but that had no bearing on one's practical life. The insistence that Torah was decidedly not some mental exercise, but rather a way of life, a *Torat Chaim*, was infuriating. "Foolish Jews!" they exclaimed, "Don't you see the beauty? The depth of what you have? Why muddle the intellectual beauty of the Torah by insisting on bringing into the minutiae of physical life?"

Precisely because it was such a beautiful inheritance, it should be left solely to the act of study and contemplation, they argued. To do anything more was to cheapen it.

That is why they, according to this account recorded in the tradition, insisted that the Jews denounce their connection to God on the horns of their oxen. "Worship your God all you want," they reasoned, "but do it in synagogue. The field is no place for meaning and Godliness."

What the Greeks failed to realize, and what the Maccabees were willing to go to war over, was that this duality between the spiritual and the physical, the divine and the mundane, only exists in our eyes. To the God, who created both the spiritual and the physical, they are both just different ways in which God decides to express God's self.

The Torah is not simply a philosophical exercise; it is a source of wisdom which helps us see the holiness inherent in the mundane, and reveal it for the world to see.

This is the reason we decided to publish Torah && Tech, to help us Torah techies stay grounded. So that the morals and ethics of Torah will guide the technology we create and that the lights we kindle this Chanukah stay lit throughout the year, in every line of code we write.

GENESIS - BERESHIT

1

BERESHIT
OBJECT-ORIENTED GENESIS

Zoë Lang

*Z*oë *Lang lives in Cambridge, MA, and is currently the Systems Implementation Consultant at Maimonides School in Brookline, MA. She is on the leadership team of the Cambridge-Somerville Open Beit Midrash and gabbait for the Orthodox Minyan at Harvard Hillel. She has worked as a software engineer in adtech and edtech.*

Of all the chapters in the Torah, few lend themselves so well to object-oriented programming (OOP) as the opening of Genesis.

A quick refresher: OOP organizes code into objects, also referred to as 'classes,' that share similar attributes. By identifying these larger patterns, objects can be invoked in a consistent way, resulting in code that is clear and easy to follow. If you think of almost any social media site, you can probably easily identify its most prevalent classes, such as users, posts, comments, and some form of 'likes.' Classes can have related subclasses that have similar but slightly different traits (maybe 'likes' can also be other reactions, such as sadness and laughter). They can even be abstract, meaning that they cannot be directly instantiated, but provide an initial blueprint for the concrete object.

2

For example, if you were creating a program to classify types of dogs, you might have an abstract Dog class to delineate general features that apply to the entire species. A concrete version of your Dog class might be a Poodle class. Then your Poodle class might have three subclasses representing different breed sizes: toy, miniature, and standard.

From an OOP perspective, Genesis 1 contains numerous objects that are instantiated over the first six days, such as Day, Night, Sky, Earth, Sea. If we want to do further refactoring, we can add some methods too. For instance, both Earth and Sea share the task of bringing forth creatures -- a method calling a Creature class would make sense, along with some subclasses for specific types (such as 'GreatSeaMonster' and 'CreepingCreatureThatCreepeth,' to borrow from an older translation). Creatures are then blessed, which could be another method for this object.

Nonetheless, Genesis 1 presents some challenges for both commentators and programmers alike. One of the best known of these is how to classify light.

In the third verse, God famously declares, 'Let there be light!' and it is created, then separated from darkness -- light is called Day and darkness is called Night. But the Sun, Moon, and Stars are not created until the fourth day; on a related point, what is providing the light of Day if the Sun does not yet exist?

Bereshit Rabbah has two explanations. Perhaps the initial creation of objects is in a more fluid state, and they only become solidified later on -- or, to put this in OOP terms, Light originates as an abstract class but needs to manifest itself concretely as Day, Sun, Moon, and Star.

The second explanation: the light that was created at the beginning was more powerful than any that exists today, and it is currently hidden; in the future, it will be visible to the righteous. However, God did allow it to appear once more: on the first Shabbat, this light was present throughout, meaning that Adam and Eve were able to see it during their time in the Garden of Eden. This also explains why, unlike for the previous six days, there is no account of morning and evening on the inaugural day of rest (Genesis 2:2). It is harder to come up with an analogy to code here, as ideally, a good codebase will excise anything that goes unused. Perhaps, for now, this primordial light is tucked away in version control, waiting for a time when this class can be restored.

One of the most amazing aspects of the Torah is how many insights it contains, although it can be easy to read only at a superficial level, we turn back to texts we think we already know. While OOP may not work for all chapters (and sometimes you need procedural code anyway!), attempting to understand the text from a novel vantage point can be one way of ensuring we remain assiduous readers as we revisit what is already familiar. Shabbat shalom!

2

NOACH
THE FLOODS OF EMAIL

This week we will be reading Parshat Noach in synagogue and recounting the experience of the great deluge brought upon humanity by God. We will read of the righteousness of Noach and the sins of the generation that he found himself in. What was that generation's undoing? The Torah tells us that the earth becomes filled with violence. Humanity had become violent people. We lost sight of the respect due to each other as created in the image of God.

Violence can take many forms, and it is not only physical. We do violence to each other when we insult each other. We do violence to each other when we humiliate each other. We do violence to each other when we violate the basic tenets of human decency and mutuality. Have we become less violent? Have we grown in our love for our fellow person? Do we know how to disagree agreeably? Have we learned the lessons of that generation that Noach was rescued from?

Perhaps one of the most destructive forms of contemporary violence in our society from person to person is the email. Emails are depersonalized and detached from the actual people who we are addressing. With a click of the "send" button, we can bring about massive insult and hurt, and we can include a limitless amount of people in the email thread. Indeed, once we hit the send button, we have lost control over that email. Who forwards it to whom? Who prints it out and shares it with others?

When we are angry or frustrated, the modern person finds themselves in front of the computer screen, hitting the keys on the keyboard writing fast and furiously. Before a breath can be taken, that person sends it into the electronic world. In the real world, we modulate our anger because we are in communication with each other. We read body language and verbal reactions. We can feel shame and regret over horrific words said. That's all in the real world, but that does not exist in the electronic world.

As we read the story of Noach and of the epidemic of violence that swept through his society, let us commit ourselves to email less and call more and to re-engage the quickly disappearing art of conversation. Let us commit fewer acts of violence to each other, whether physical or with our words, and think before hitting that send button.

3

LECH LECHA
ARE YOU A 10X ENGINEER?

Hiring managers, recruiters, CFOs, founders, all of them are on the search for the mythical "10x Engineer."

"Hire a 10x Engineer," common wisdom goes, "and your codebase will practically write itself!"

But what is a 10x Engineer? And how do you find one? Is it the bearded dude sitting alone in the corner of the office surrounded by cans of energy drinks and banging away in Vim on his mechanical keyboard?

In this week's Torah portion, Lech Lecha, we learn all about Avraham, the founder of monotheistic Judaism.

Avraham accomplished a lot in his life. According to the Medrash, he smashed his father's Idols, was thrown into a fiery furnace by a threatened King Nimrod, and was given ten tests by God and passed them all. It is no surprise then that God referred to Avraham as "Avraham Ohavi" (Avraham, the one who loves me).

Yet, if you wanted to know why it was that God chose Avraham to father the Jewish nation, the Torah doesn't mention any of that.

כי ידעתיו למען אשר יצוה את־בניו ואת־ביתו אחריו ושמרו דרך יהוה לעשות צדקה ומשפט

"I have known him because he commands his children and his household after him, that they should keep the way of God to perform righteousness and justice…"(Bereshis 18:19)

What made Avraham stand out from the others wasn't his own righteousness; Noah was righteous as well. But, while Noah was content with being righteous and wasn't too worried about the rest of his generation, Avraham traveled the world to spread knowledge of God and "righteousness and justice" to everyone.

In the words of Rabbi Yosef Yitzchak of Lubavitch:

> "We find that God's love for our father, Avraham, was mainly because '…he will command his children and his household.' All of Avraham's work in service of God, and in the tests to which God subjected him, cannot be compared to his teaching others and bringing them closer to God, i.e., to his bringing merit to others."

> (Igrot Kodesh vol. 6 page 91, brought in Hayom Yom for 8 Tammuz)

So what is a 10x Engineer? A 10x Engineer is not an engineer who writes 10x more code than other engineers. A real 10x Engineer is an engineer who can boost the productivity of their team 10x. An engineer who mentors, uplifts, shares their knowledge, and ultimately brings ten other engineers up to their level.

So, please don't go chasing unicorns. Instead, be like Avraham, and no need to stop at 10x Engineer! Be a 10x Jew and a 10x human being!

4

VAYERA
THE HEADPHONES CAN COME OFF
SOMETIMES

How many times have you been in the midst of a particularly challenging project, and someone has tapped your shoulder to get your attention? You may be near the final touches of a new algorithm, or working on a complex design, or constructing a SQL query that is just running on too many lines, and this person breaks your attention. What do you do?

I know I have certainly been there. In one of my early software positions, there was one very well-meaning non-technical colleague who loved to schmooze. There were times where I definitely found it a frustrating experience.

In this week's Torah portion of Vayera, we find Abraham also in the midst of his own challenging moment. The portion opens with him sitting outside his tent during the heat of the day. The classic commentators teach that he was recovering from his own circumcision. He was recuperating from what, at his age, was a significant surgical procedure. Yet, it was precisely at that moment that he sees three travelers on the road.

What does he do?

Abraham runs after the strangers. He insists that they rest for a while. He arranges an elaborate meal for them. He applies his world-renowned hospitality to these guests, even amidst his recuperation. He saw people that needed help, and despite his own challenges, he took the time to help them.

Yes, there are times where we must keep the headphones on. But, sometimes, we can take them off and address the person in front of us. If Abraham could help strangers after surgery, we can summon the courage to help those amidst an algorithmic challenge.

5

CHAYEI SARA
ON ARTIFICIAL (UN)INTELLIGENCE

This week's Torah && Tech will take a different format than usual. Where most weeks we introduce a point from the week's Torah portion and try to learn a lesson from it to our tech lives, this week, I thought I would share some thoughts I had.

It will be more of a "stream of consciousness" thing and will probably read more like a Talmudic debate than a sermon. You might end up with more questions than answers (and that's a good thing). The intent is to start a conversation. If you would like to share your thoughts on the issues raised, please feel free to reach out to Ben or me, we'd love to talk to you!

-Yechiel

Recently, an uproar erupted when it turned out that a certain company in the business of giving out credit cards was discovered to have routinely extended less credit (sometimes 20x less) to women than to men under the same financial circumstances.

When called out on their discriminatory practices, the company defended itself by saying that all of the decisions were being made by an algorithm, and could therefore not be biased.

In this case, the effect of the biased algorithm was financial. Yet, algorithms have been called upon to make much more serious decisions, such as who should be let out on bail and for how much, and how healthcare should be administered; these are decisions with potential life-and-death implications.

A while ago, I saw a halachic discussion on whether we can hold an artificial intelligence (AI) liable for its actions.

Currently, that conversation is purely theoretical, as that would require AI to have an understanding of right and wrong that is way beyond anything we have at the time. But recent events did get me thinking about a slightly related question; if someone programs an AI to make some decision, and the AI causes harm, to what extent is the one who deployed the AI liable for the AI's actions.

Put in simpler terms, is saying "it wasn't me, that decision was made by the algorithm" a valid defense?

Finding the Torah view on this question is understandably challenging; the Torah, after all, doesn't speak of computers. We will have to get creative and see if we can find an analogous situation.

One approach we might take is to consider the AI as a Shliach, or an agent, of the person who deployed it.

In Halachah, a person can appoint a Shliach to do an action on their behalf, and the activity will be attributed to the appointer (the Meshaleach). For example, if you appoint someone to sell something on your behalf, the sale is attributed to you as if you executed the transaction.

What if the person appointed the Shliach to do something wrong (e.g., to steal something)?

In such a case, the Talmud rules that אין שליח לדבר עבירה (the concept of a Shliach doesn't apply where sin is involved).

In other words, the Shliach is expected to refuse the problematic job. Should they go ahead with it anyway, they are liable for committing the sin.

Trying to apply this rule to AI, though, leads to some problems. The reasoning behind the ruling that Shlichut doesn't apply when sin is involved is because the Shliach is expected to use their moral judgment and refuse the Shlichut. This requires that the Shliach have a sense and knowledge of right and wrong, as well as the autonomy to make their own decisions, two things AI is nowhere near achieving, as mentioned earlier.

It would seem that using AI would be similar to using a tool. Similar to when a person kills someone using an arrow, they can't defend themselves by saying, "it wasn't me, it was the arrow," using an AI might be the same thing.

You might argue that AI is different than an arrow in that once you deploy the AI, you "lose control" over it. The algorithms behind AI are a "black box," even the programmers who programmed and trained the AI are unable to know why it makes the decisions it makes.

So unlike an arrow where there is a direct causal relationship between the act of shooting the arrow and the victim getting hurt, the causal link in AI is not so clear-cut.

But then again, it seems like the Talmud discusses a case that might be analogous here.

Regarding a case where a person lit a fire on their property and the fire got out of control and damaged a neighboring property, the Talmud says the following:

> "We have learned that Rabbi Yochanan said: [he is liable for] his fire just as [he is liable for] his arrow."

A closer look at the reasoning behind Rabbi Yochanan's ruling, however, reveals a crucial difference. The reason you are liable for your fire spreading is that fire spreading is a predictable consequence of lighting a fire. If your fire spread due to unusually strong wind, for example, then you would not be liable because the spread of the fire could not have been predicted.

One can argue, the fact that the AI made its own decision could not be predicted, even by those who programmed the AI and wrote the algorithms behind it. This would mean that the AI has some sort of agency here. Maybe not enough agency to hold the AI liable, but perhaps just enough to exculpate those who deployed it (as long as they are unaware that the AI is making faulty decisions).

Is there something between the full agency of a Shliach and the complete lack of agency of a tool/fire?

Let's look at another passage:

> "One who sends fire in the hands of a child or someone who is mentally impaired is not liable by the laws of man but is liable by the laws of Heaven."

The idea that someone can be "not liable by the laws of man but liable by the laws of Heaven" is often used in the Talmud to refer to actions that are technically legal, but still unethical. So while the court can't prosecute a person for a fire started by a child, it is still morally and ethically wrong to do so.

So perhaps that is how we can classify AI? Like a child who has enough agency to make decisions, but not enough to distinguish right from wrong? Are companies hiding behind black-box algorithms technically legal, but morally questionable?

As I said in the beginning, I don't know the answer to these questions, but I do hope we can start a discussion because the days where such questions were the realm of theoretical philosophers are coming to an end faster than we think!

6

VAYETZEI
MAINTAINING THE PROPER
HEADSPACE

In the beginning of Parshat Vayetze, we read about how Yaakov leaves his parents' home in Be'er Sheva and heads to his uncle Lavan in Charan, where he will spend the next 21 years working as a shepherd, marrying, and building a family.

At the beginning of his journey, the Torah describes how the sun set just as he reached Mount Moriah, the future site of Jerusalem and the Holy Temple.

As Yaakov prepared to go to sleep, the Torah describes how he took a few stones and put them around his head for protection before lying down.

A seemingly obvious question that comes up is, why did Yaakov find it necessary to protect only his head? What about the rest of his body? Wasn't he worried about that?

The Lubavitcher Rebbe OBM gave an interesting answer.

Yaakov was leaving his parents' home where he was able to spend most of his time "sitting in the tents" of study. Indeed, the Medrash tells us that Yaakov had spent the previous 14 years learning in the Yeshivah of Shem and Ever.

Now Yaakov was getting ready to leave it all behind and enter the workforce, and as preparation, he was teaching us a valuable lesson.

The verse in Tehillim says:

"יגיע כפיך כי תאכל, אשריך וטוב לך"

If you eat the toil of your hands, you are praiseworthy, and it is good for you.

(Tehillim 128:2)

Working and making a living should involve only your hands, your head should remain free for the things that are important in life; your relationship with God, your family, and yourself.

This advice can be harder for those of us who's work specifically requires that we involve our heads, and that makes it that much more critical to maintain proper boundaries between our work lives and our personal and spiritual lives!

A story is told about Rabbi Shalom Ber of Lubavitch (the 5th Chabad Rebbe), who once noticed that one of his disciples looked preoccupied. When Rabbi Shalom Ber asked the Chassid what the matter was, the Chassid started telling him about the problems he was experiencing with the galoshes factory that he owned.

The Rebbe looked at him and said, "I have heard of people putting their feet in galoshes, but to put your head in galoshes...?"

So let's try and take a lesson from our grandfather Yaakov and, even as we go out in the world and create amazing things, let's make sure to maintain our safe-space around the things that matter most.

7

VAYISHLACH
ALL IN THE DETAILS

(Dedicated in loving memory of Dan Pesach Ben Baruch Shmuel Hakohen Greenberg)

It's All in the Details

Last week, after the weekly Torah && Tech newsletter went out, we were saddened to hear about the passing of Ben's father, Dan Pesach ben Baruch Shmuel Hakohen Greenberg. The T&&T team would like to extend our warmest wishes to the Greenberg family, and may they be comforted among the mourners of Jerusalem. This week's issue is dedicated to his memory.

In this week's Torah portion, Parshat Vayishlach, we read how Yaakov, who, after spending 21 years in Charan in the home of his uncle Lavan, was now on his way home to the Land of Israel.

On the way, the Torah describes how Yaakov took his family and belongings across the Yabok river. The Torah then describes how "Yaakov was left alone, and a man wrestled with him until dawn." (Bereshis 32:25)

Why was Yaakov on his own? The Talmud (Chulin 91a) says that Yaakov went back across the river in the middle of the night to retrieve some "small jugs" that were left behind.

From this, the Talmud learns the value of even little, seemingly insignificant things. Even a few "little jugs" weren't too insignificant for him, and he refused to leave them behind.

As developers, we are all too painfully aware of the importance of the little things.

How many times have you spent time hunting down a bug that turned out to be a misplaced closing bracket or a forgotten capitalization?

What we need to keep in mind is that the "little things" count in our interpersonal relationships, as well as our interactions with our teammates and co-workers.

We have no idea what a "good morning" said with a smile and a genuine interest in the well-being of a teammate who seems to be burning out can accomplish.

Let us try our best to pay attention to the little things, the ones that make the biggest difference!

8

VAYESHEV
FAMILY: IT'S COMPLICATED!

Raise your hand if you have ever dealt with complicated family dynamics? How about complicated workplace dynamics? Complicated friend dynamics?

I imagine just about everyone can relate to what is essentially the challenge of living our lives in the midst of other people. Sometimes we may just want to find a cave and retreat from everyone. Alas, that is not a possibility for most of us.

We must find a way to make do in this world. The threads of relationships afford us many blessings while also being threaded with challenges. There is not much in this life that is wholly good or wholly negative. Often, they are one and the same.

One does not need to look farther than this week's Torah portion to discover that this has been true and part of the human condition for as long as there have been human societies. Deceit. Jealousy. Attempted murder. Slavery. Lust. They all find their way into the portion of Vayeshev.

Within all that tumult and havoc, we also encounter perseverance and determination. Perhaps, learning to thrive in the reality of life is one of the most important lessons we can glean this week from the Torah. If we only could live inside our code, if we are developers, life could be a lot easier, but it would lack the richness and opportunity for real personal growth.

Joseph, the character who so much jealousy, rage, and lust is thrust upon, time and time again finds a way to turn the lemons into lemonade. Perhaps our work this week is to find ways to flex our perseverance muscles in the challenges we face and learn from Joseph? It is certainly not an easy task, but the reward to our life and the daily experience in the communities, families, and workplaces we call home may very well make it worthwhile.

9

VAYIGASH
VIRTUAL IDENTITIES

"And Joseph said to his brothers: 'I am Joseph, is our father still alive?'"

(Genesis 45:3)

"Am I my avatar's keeper?"

(Dr. Sherry Turkle, *Alone Together*)

This week's Torah portion of Vayigash places us at the climax of the eventful moment when Joseph reveals himself to his brothers. Separated by time and place, their relationship torn asunder by distrust, jealousy, and betrayal, they now must bridge those divides to reunite their family and bring peace to their elderly father. It is no coincidence that at the moment of revelation, the first words uttered by Joseph are inquiring as to their father's health. It is often the deterioration of a family matriarch or patriarch that can bring siblings together.

Focusing in on the events unfolding here, in this moment, we are confronted with the lifting off of the mask of anonymity and concealment that Joseph wore in front of his siblings. This was both a real literal mask, the decorations and adornments of ancient Egyptian aristocracy, and it was the metaphorical mask of emotional distancing.

At this charged moment, the first question asked is one that was already answered. When the brothers introduce themselves to the viceroy of Egypt, who unbeknownst to them was actually their brother, they declare that they have an elderly father awaiting them back home in the Land of Israel. Joseph already knew about the condition of their father, so why ask again? And, why ask again, precisely at this powerful moment?

Many of the classical commentaries address this question. They all point to the most modern of technological challenges: How to maintain integrity in the face of anonymity?

The 16th-century sage Rabbi Ephraim Luntschitz, known for his popular Torah commentary entitled Kli Yakar, addressed this clearly when he explained that Joseph thought that perhaps everything the brothers stated prior to his revelation of identity was in order to arouse sympathy from a ruling leader for a band of migrants seeking relief. Yet, now that they knew his other identity, things might be different.

In other words, who we are and who we are perceived to be is different depending on the identities we project outwards. What is your identity on Twitter? What is your identity on Stack Overflow? Do you have different identities for different Slack networks? How do you present at work and how do you present at home? Would someone think to ask you the same question twice and expect two different answers depending on circumstance and place?

While we often seek a bringing together of our disparate selves, there is also a time and a place for the divergence of self into smaller component parts. It made sense to Joseph, initially, to conceal one part of himself (his familial connection) and project exclusively another part (his Egyptian aristocratic identity). At a later point in our Torah portion this week, the time came to share both identities. Joseph demonstrated a keen sense of self-awareness as to which aspect of his identity to deploy and when. Do we possess the same sense of self-awareness? If not, how can we expand our own understanding of ourselves?

In our time, we have a multitude of masks we can wear at any single moment. Amidst all the masks -- all of the identities we possess -- can we still keep track of our actual face? Do we possess the awareness of self enough to know when we are wearing one of our identities, which one it is, and when it is time to lift that mask? Joseph, through his challenging and complicated personal story, presents a path forward. We can, if we choose so, be our avatar's keeper.

10

VAYECHI
THE CASE FOR ZEVULUN

"Zebulun will dwell on the coast of the seas; he [will be] at the harbor of the ships, and his boundary will be at Zidon.

Yissachar is a bony donkey, lying between the boundaries."

(Genesis 14-49:13)

"Zebulun and Issachar entered into a partnership [with the following agreement]: Zebulun would dwell at the seashore and go out in ships, to trade and make a profit. He would thereby provide food for Issachar, and they would sit and occupy themselves with the study of Torah. Consequently, Moses mentioned Zebulun before Issachar [even though the latter was the elder of the two], because Issachar's Torah came through Zebulun."

(Rashi Deuteronomy 33:18)

Throughout the ages, there existed a partnership among the Jewish people. The "Zevulun"s, who engaged in business and trade, would support the "Yissachar"s, the select few who could devote their lives to the study of Torah.

Conventional thinking has always been that Yissachar is what we should all strive to be, while the Zevuluns only play a supporting role. The Zevuluns are "stuck" dealing with the materialism of the business world while the Yissachars are free to pursue the greatest spiritual heights.

The Ba'al HaTanya offers another perspective.

The purpose of creation, says the Ba'al HaTanya, is not to attain spiritual heights; the highest of spiritual heights are still an infinite descent to God. Rather, the purpose of creation is, as the Midrash says, that God "desired to have an abode in the lower realms."

In other words, our purpose in life should not be to depart this physical world in a fiery chariot of spirituality. Instead, our purpose is to bring spirituality into the world. Toward that end, God gave us the Torah; to teach us how we can draw spirituality into a physical world.

When we look at it from this perspective, it's easy to see why Zevulun is the main actor in this play. Zevulun is the one who goes into the world, who meets people, who creates, who elevates, who is a "light unto the nations."

That isn't to say that Yissachar doesn't have an important role; it is Yissachar, after all, who teaches and inspires Zevulun to go out and bring the light of Torah into the world. But, at the end of the day, it is Zevulun who gets the job done.

Even though Yissachar was the older one among Yaakov's children (because it is Yissachar who teaches, leads, and inspires), it was Zevulun who was mentioned first in the blessings of Yaakov and Moshe. To quote Rashi, "Yissachar's Torah came through Zevulun," which can be understood to mean that Yissachar's Torah was actualized through the actions of Zevulun.

In short, from a Torah perspective, when we make sure the technology we create and the interactions we have at work conform to our Torah values, we are not just making "the best" of a situation, we are fulfilling God's purpose of giving the Torah in the first place!

To conclude with a story:

The Chassid Reb Elyeh Abeler was a simple man. Once, when he came to visit his Rebbe, the Rebbe Maharash of Lubavitch, the Rebbe said to him: "Elyeh, I envy you. You travel to various markets and fairs for business; you meet many people. Sometimes, in the middle of a business transaction, you mention a Dvar Torah, a story from the Ein Yakov, etc., you arouse interest in studying Torah. This causes joy On High, and the Almighty rewards such 'business' with the blessings for children, health, and livelihood; the larger the fair, the more work there is, and the greater is the livelihood earned."

EXODUS - SHEMOT

1

SHEMOT
WHOSE RESPONSIBILITY IS IT?

"Now the king of Egypt spoke to the Hebrew midwives, one who was named Shifrah, and the second, who was named Puah. And he said, 'When you deliver the Hebrew women, and you see on the birthstool, if it is a son, you shall put him to death, but if it is a daughter, she may live.'"

(Exodus 16-1:15)

The declaration from the Pharaoh is unthinkable and utterly devastating. How can one grapple with such an edict of unspeakable violence? Confronted with this order, Shifrah and Puah decide enough is enough. The following verses share that these midwives did not do as he commanded, but rather continued their sacred work in bringing new life into the world.

The Talmud (Sotah 11b) in discussing this episode relates that not only did they deliver the newborn babies into life, regardless of their gender, but they also went a step further and made sure that they had the necessary nourishment and supplies to thrive in life. They not only refused to obey an immoral directive, but they actively worked against it.

While, in our industry, we thankfully do not often face clear life or death scenarios, we do often encounter murky areas of moral responsibility. With news stories continuing to evolve on how some companies broke the common trust with their users' data, serving as just one example among many. What is our moral responsibility when faced with a directive that we feel crosses a moral or ethical line? Are we collectively liable if we are only following an order from management?

These questions and more are at the heart of the encounter of the midwives and the Pharaoh. Next time you find yourself confronted with a morally challenging work ticket on Jira, perhaps ask yourself: What would the midwives Shifra and Puah do?

2

BO
WE'RE FREE... LET'S MAKE
A CALENDAR!

"This month shall mark the beginning of the months; it shall be the first of the months of the year for you."

(Exodus 12:2)

What is the first thing you do when you wake up in the morning? What is the last thing you do before you fall asleep each night? If you are like most people, you are probably staring at your smartphone screen.

"Let me just check Twitter one last time."

"I wonder what's new on Facebook this morning?"

"Who emailed me overnight?"

If you have ever tried to leave your house for the day without your phone, you know that feeling of displacement and loss. We are so tethered to our untethered devices, and much commentary has been written on it. All the tech that surrounds us and pervades our life is supposed to make things simpler and our life less cumbersome. Oftentimes, we end up feeling more encumbered and burdened. Do we own our tech, or does our tech own us?

In this week's Parsha, we encounter the first mitzvah -- the first commandment -- given to the newly born Nation of Israel. A nation of slaves, on the cusp of freedom, are given their first collective responsibility. What is this Divine directive? What commandment could help shape their newfound destiny as free people and a free nation? Well, nothing less than the creation of a calendar.

A calendar? Really? In this momentous and auspicious time, the first mitzvah is a calendar? Doesn't that seem a bit too technical and a bit devoid of spiritual power and purpose?

The reality is that nothing could be more appropriate than the mitzvah to establish a calendar at this moment. In declaring the mitzvah, the Torah states: "... it shall be the first of the months for you." The 15th-century rabbi, Ovadia Seforno, teaches that the inclusion of the "for you" was very deliberate. Up until this moment, their time was not theirs. It belonged to their masters. They were slaves, not in possession of the most basic element of a free person: the right to choose their own time.

A calendar is a technology. It is one of the oldest and most continuously used technologies in human civilization. When we exercise control over it and make it for ourselves, it allows us to express our freedom. Conversely, our calendars can overpower us. We can become enslaved to our agendas, just like we can become enslaved to our smartphones.

Maybe this week we can find opportunities to make our technologies subordinate to us and not the other way around. Check your phone first thing when you wake up? Maybe deliberately not do that for a few days. Feel overwhelmed by your schedule? Maybe put some personal time into your weekly calendar and reclaim your time. Try to reclaim some of your life and exercise your freedom and see how it goes.

3

BESHALACH
KEEP CALM... AND GO FOR A WALK

"And God said to Moses, Why do you cry out to Me?
Speak to the children of Israel and let them go forward."
(Exodus 14:15)

In this week's Torah reading, we read about the story of the splitting of the sea. We are all familiar with the setup. The newly freed Israelites were traveling through the desert when they found themselves between a rock and a hard place, or, in this case, between the mighty Egyptian army in hot pursuit behind them, and a stormy sea in front of them.

The Medrash describes an intense scene that took place at this point: "As they stood at the shore of the sea," the Midrash says, "the people of Israel split into four factions."

One faction said: "Let us cast ourselves into the sea." A second faction said, "Let us return to Egypt." A third said, "Let us wage war against the Egyptians." A fourth said, "Let us pray to God."

After addressing the concerns of each of these four factions individually, God spoke to Moses and said, "Why do you cry out to Me? Speak to the children of Israel and let them go forward."

Moses passed on the message, the Children of Israel pressed forward, and then... never mind, I won't spoil it for you, you'll hear about it in Shul.

Many times in our lives we face challenges that seem to get in the way of us doing what we need to do. Those challenges can take the form of obstacles in our path (the Red Sea), or distractions from the outside (the Egyptian army).

Many times our reaction to these challenges is to start overthinking things and brainstorming solutions (or even entertaining thoughts of giving up). When the reality is that if we just kept going and doing what we have to do, we would discover the challenges are more a product of our making than anything else.

The Jewish people leaving Egypt were given a task: "When you take the people out of Egypt, you will worship God on this mountain." They were on a mission to get to Mount Sinai to complete their transformation into a nation. Any obstacles in their path should have been immaterial. There's an ocean in the way? Who cares? Are the Egyptians chasing behind you? Why are you even looking? "Speak to the children of Israel and let them go forward."

I often thought of this Medrash in my early days of learning programming. Back then, I would often run into problems that seemed impossible.

Who of us hasn't faced those moments when you are staring at the screen for an hour, two, six, and nothing seems to work?

Sometimes I would just be banging away at the keyboard trying to brute-force my way to a solution ("let us fight them"). Other times I would be tempted to give up on the problem and find something else to work on ("return to Egypt") or drop out of tech altogether ("throw ourselves into the sea"). Perhaps I should leave my pride behind and ping my instructor for the umpteenth time ("Pray to God")?

More often than not, in those days, it turned out I was completely overthinking things. The simple, straightforward solution I had discarded in the beginning because "it can't be THAT simple" was actually the right way to go about it.

Sometimes the answer WAS to keep "going forward" and not try to overthink, overcomplicate, and see problems where there were none.

Then there are the times where you tried all the solutions; you looked at all the answers from Stack Overflow, you spent hours printing debug statements, and still, the puzzle refuses to crack.

At times like those, a more literal approach may be necessary. "Go forward," step away from your computer and just go for a walk. It's amazing what clearing your mind and looking at the problem with a fresh pair of eyes and a more open mind can do.

Sometimes, the road to the promised land starts with a walk.

4

YITRO
YOU CANNOT DO IT ALONE

"... For the task is too heavy for you; you cannot do it alone."
(Exodus 18:18)

Between daily standup meetings, weekly deliverables, Kanban boards, and more, we can feel like we have no time to accomplish what we need to do. There are just not enough hours in the day! If you think you are busy though, just look at the workload Moses faced every day in the wilderness.

Hundreds of thousands of people. Hundreds of thousands of anxious, tired, scared, exhilarated, and recently freed slaves. They all needed answers, and they could not wait. When you need answers in the midst of the wilderness and in the midst of the Exodus, who do you turn to? Moses, of course! That left for a very overburdened leader.

In fact, as soon as his father-in-law Jethro sees him, it is the first thing he notices about his son-in-law; His absolutely unmanageable schedule. It is his father-in-law who suggests the change. A change that was not only absolutely integral to the life of Moses, but also to our own work-life balance as well.

Delegate! Find colleagues with who you can collaborate. Find people who can share in your burdens and in your success.

The work is often too great to do it alone. Furthermore, even if we could accomplish it all on our own, we are better off when we invite others to join us. The work becomes enriched by the introduction of multiple perspectives and approaches.

Friends, let's try and hear Jethro's advice this weekend as if he was speaking to each one of us. We shouldn't and we can't do it all alone.

5

MISHPATIM
AM I MY CODE'S KEEPER?

"If one leads his animals into a field or a vineyard, or lets his animal loose and it eats in another's field, the best of his field or the best of his vineyard he shall pay. If a fire goes forth and finds thorns, and a stack of grain or standing grain or the field be consumed, the one who ignited the fire shall surely pay." (Exodus 5-22:4)

In the first Torah portion after the momentous revelation at Mount Sinai, the Torah is surprisingly mundane - three chapters of mostly civil law. Laws of how to deal with lost property, monetary damage, financial matters, the extent of responsibility for borrowed or rented objects, monetary disputes, liability for injury caused by one's self or property, and so on.

Many of those laws may seem like they don't apply today to our urban lifestyle. Do we really need to worry about our cattle grazing in our neighbor's property? (Our dogs maybe, but, as the internet likes to say, they're good dogs, Brent).

A look in Talmud, however, will show that the two verses quoted above have generated pages upon pages of debate, and the modern applications of those debates are still being debated today in Yeshivahs and study halls worldwide.

Earlier this week, the tech world was rocked by the news that a popular app used by millions to communicate contained a vulnerability that allowed it to be used to spy on its users. Then, perhaps more maliciously, one of the largest social media platforms was reported to have paid teenagers to install a VPN that handed over virtually all of their data to the social media giant.

As we write software and "let it loose" into the world, we have to ask ourselves: To what extent are we responsible for the consequences of the software we write? What if those consequences will only manifest further down the line? If an algorithm we wrote to help locate a cellphone ends up being used to spy or track people, is that on us?

There are no easy answers to those questions. They aren't even easy questions to ask. What we do know is that we can't afford not to ask them.

These are discussions that have been going on for thousands of years and still going on today. These are conversations we have to have with ourselves as developers, as a team, as a company, and as an industry as a whole.

The answers won't come easy, but hopefully, we'll learn a lot just from asking the questions.

6

TERUMAH
THE MENTORSHIP RELATIONSHIP

"Moses was puzzled by it..." (Rashi on Exodus 25:31)

This week's Torah portion reads like an architectural blueprint. We work our way through the detailed plans for the construction of the Mishkan, the Tabernacle, in each section of the portion. God elaborates on each instrument of the Tabernacle to Moses, step by step. It is a true "senior to junior" mentorship. In many ways, not so dissimilar from the ways we interact with our colleagues at work, whether we are the "senior" or the "junior."

In fact, if we take a step back, it can be helpful to think about the ways in which we engage in the act of mentorship. How do you mentor? How do you receive mentorship? The Harvard professors, Ronald Heifetz and Marty Linsky, in their book "*Leadership on the Line*", argue that being a leader is less a noun than it is a verb. In other words, it is a deliberate action. You act with leadership. Every person, regardless of their formal status or title, can exercise leadership throughout their lives.

So too, everyone cannot only be mentored but can also exercise mentorship.

Let's return to the God and Moses relationship as outlined in this week's Torah portion. In the midst of the instructions, we find a moment where Moses is confused. The manual for the building of the menorah is particularly detailed, and everything has to be just right. Rashi, the great medieval exegete, notices this confusion and teaches that God noticed it as well. As a result of that confusion, God decides to be more explicit and not let Moses hit his head against the wall in exasperation, and just simply explain it to him.

Mentorship is about helping those you mentor learn the tools needed to succeed. Yet, at times, what will help those we mentor succeed, is being shown the answer when the situation is entirely too complex and too challenging. If we are the ones being mentored, it is also knowing the times when it is okay to ask for that extra help. The road to success is paved with a lot of banging on the keyboard in frustration as part of the learning experience, but it is also knowing when to give that extra bit of help and when to receive it.

In their relationship, Moses and God show us the nuanced ways of true mentorship.

7

TETZAVEH
WHAT'S THE POINT OF POINTERS?

This week's Torah portion, Tetzaveh, has the unusual distinction of being the only portion in the Torah, starting with his birth in Shemot, where Moshe's name is not mentioned even once.

The 13th-century commentary, the Ba'al Haturim, gives an interesting explanation. After the Jews sinned with the Golden Calf, Moshe was begging God to forgive them, saying, "Now, if You will forgive their sin [well and good]; but if not, erase me from the book which You have written!" (Shemot 32:32).

The Ba'al Haturim goes on to explain: "Although this exclamation was conditional upon God's refusal to forgive the people for the sin of the Calf, the curse of a scholar is fulfilled in some way, even when made conditionally." So even though God ended up forgiving the Jewish people and didn't have to go through with Moshe's "threat," he nevertheless went through with it to a smaller extent by removing Moses's name from one portion of the Torah.

The Lubavitcher Rebbe OBM used to point out a, seemingly ironic, fact. Even though God removed Moshe's name from the Torah portion, Moshe himself was not. The portion of Tetzaveh contains numerous references to Moshe in the first person, starting with the very first verse, "And you shall instruct the Jewish People..."

In a certain sense, the Rebbe explained, Moshe's presence in Tetzaveh is actually felt much stronger. Where a name is external, only used by others to address us, "you" refers to your essence. In other words, in "erasing" Moshe's name from the Torah, God wrote in Moshe "himself."

Though it may seem counter to Moshe's intent, it was actually pretty fitting. In Moshe's declaration "erase me from the book which You have written," Moshe revealed his true essence as the ultimate leader, willing to put his life's work on the line to save his people.

What does all of this have to do with tech? I'm (Yechiel) glad you asked (and I would like to apologize if this part gets more technical than previous newsletters).

Being relatively new to the Go language, one of the features that confused me a lot at first was the idea of pointers. Coming from Ruby and Python, where pointers don't exist, I had a hard time wrapping my head around them and why they were necessary. Why is a pointer to a location in memory where a value is stored any different than a variable that holds that value? I believe the idea above, about the difference between Moshe's name and Moshe himself, can help us understand.

Consider the following code sample:

```
func increment(num int) {
  num+=1
}

func main(){
  var x = 1
  increment(x)
  println(x)
}
```

We have a function increment() that takes a number and increments it by 1. We then take a variable x, assign it to the number 1, we call increment on x and then print it. The result that gets printed is not 2 (the result of incrementing x by 1); instead, 1 gets printed.

The reason for this behavior is that when we call increment(x), the increment function makes a copy of the value of x, assigns it to a new variable in its own scope, and increments that. The original x remains unchanged.

If we wanted increment() to increment the actual value of x we could modify our code as follows:

```
func increment(num *int) {
  *num+=1
}

func main() {
  var x = 1
  increment(&x)
  println(x)
}
```

Our new increment() doesn't take a number as an argument, it takes a pointer to a number, an address in memory where that number is stored. And when increment() increments the number, it increments the number at that location in memory. Running the second code snippet gives us 2 as we expect. All because we pointed at the number itself instead of at its value.

Sometimes, in life, we have to know where to point to and what defines the mark we leave around us and what actually defines us, in our essence.

8

KI TISA
WHAT IS YOUR TECHNICAL DEBT

This week's Torah portion puts us right in the center of one of the most dramatic incidents in the entire Torah.

Moses is up on top of Mount Sinai in an unprecedented conversation with God. The people are assembled at the foot of the mountain, eagerly awaiting his return. Yet, the time begins to add up, and the people's patience begins to run short. They let their fears, anxieties, and worries overpower them. In the midst of their collective panic, they construct an idol, a Golden Calf, to help calm and soothe them. The Golden Calf reminds them of the trappings of Egypt. Even though they led a bitter and enslaved life in Egypt, it was nonetheless the world that they knew for so long, and recreating a symbol from there brings them some form of comfort.

Moses returns from his encounter with God and brings with him a set of Divinely inscribed tablets. However, upon seeing the festivities surrounding the people's Golden Calf, Moses destroys the tablets in anger. God, too, is so angry that God swears to destroy the entire people. Moses intervenes on their behalf and successfully argues against their destruction. But, this incident is not gone forever.

Rashi, the great medieval exegete, commenting on this episode (Ex. 32:34), quotes the Talmud that declares that God will always remember this incident each time God accounts for the actions of the people. This moment becomes part of the national narrative for all time.

In other words, this single moment becomes a pivotal act of accruing technical debt. What is technical debt? Technical debt is the decisions made early on in the decision of an application that causes additional work down the road. Their shortcomings are not readily apparent at the moment of creation, but only further on, does their weight begin to be felt.

We often think of technical debt in the realm of software development.

We can structure the database this way for now, and it works, but once our data reaches a certain value, that decision will come back to challenge us. We can loop through this array using this method, and it's fine for now, but what happens when that array becomes an array of arrays, will that decision we made at the beginning of our application cause us problems at that point?

Technical debt, more broadly, can also be the accrual of interpersonal, organizational, and strategic decisions we make in the present to get the job done, but come at the risk of causing us challenges later down the road. The Golden Calf felt like the right decision for the people at that emotionally charged moment. Yet, in retrospect, its implications would reverberate for generations.

This Shabbat, let's give thought to the decisions we make in our work lives, our personal lives, etc. that either make things easier in the here and now, or just make sense at the moment, but might be accruing our technical debt in the long term. Maybe there are things we can do now to help our future selves?

9

VAYAKHEL
DON'T LOSE FOCUS!

This week's Torah portion, Vayakhel, begins with Moshe gathering the Jewish people to pass over the specs he received for the project of building the Mishkan - the Tabernacle that would be God's home on earth.

Strangely enough, he seemed to take a "detour" before making the momentous announcement. In the words of the Torah:

> Moses called the whole community of the children of Israel to assemble, and he said to them: "These are the things that the Lord commanded to make. Six days work may be done, but on the seventh day you shall have sanctity, a day of complete rest..." And Moses spoke [further] to the entire community of the children of Israel, saying: "This is the word that the Lord has commanded to say: [going on to command them regarding the building of the Mishkan]" (Exodus 35:1-4)

An obvious question raised by many commentators is how does Shabbat come in here? If Moshe gathered the Jewish people to tell them about the Mishkan why didn't he focus on that?

Rashi explains that God was worried that in their great excitement to get the Mishkan built, the Jewish people might think that the work of building the Mishkan could take precedence over the Shabbat. God, therefore, instructed Moshe to tell them that no matter how important the work on the Mishkan may seem, they should not forget about the Shabbat.

There's a common attitude that is, unfortunately, prevalent in the tech world, that if your work is important to you then you'll spend all your waking hours (and some of your sleeping hours as well) immersed in it.

Many people have the attitude that "real" developers are those who stay in the office until 9 pm and then come home to work on their three side-projects, write a bi-weekly blog post, record their podcast, all between submitting talks to multiple conferences and meetups.

This week's Torah portion teaches us that no matter how important your work is - you may be building a home for God on this earth - some things come first. Taking time to connect to your soul, your family, your community cannot be pushed aside by your work to change the world.

This Shabbat, let's take a moment to reflect and reconnect. To focus on what's important and to discard what isn't. You'll find that on Monday you will be a lot more prepared to tackle your work and make the world a better place.

10

PEKUDEI
CONTINUOUS INTEGRATION
IN THE WILDERNESS

In the Torah portion for this week, Parshat Pekudei, we encounter detailed instructions regarding the Tabernacle, the Mishkan. This elaborate structure was the central focal point for the nation as they wandered in the wilderness on their way to Israel. It was the place that a person could go to feel the immanence of the Divine, and it was the place that the nation went to in order to find atonement for the entire community.

What is absolutely striking is to remember that this structure, and all of its accompanying vessels and equipment, needed to be constructed and reconstructed over and over again. Each time the people moved to their next destination in the wilderness, they were required to break it down and rebuild it again anew.

Every time a development team deploys a new release, there is always the dual feelings of anxiety and excitement that go along with it. Will it work? Did we catch the major bugs before release? Did we anticipate the edge cases? For people who don't work in software development or who don't work closely with software developers, it is hard to appreciate how big of a deal a new release can be.

In fact, it might not be so different than the feeling the leadership, including Moses, felt every time they brought forth the Tabernacle in each new build. Did they miss something? Did they tighten every loose area? Connect everything that needed to be connected? Put everything in the right place?

There is a lot of work that goes along with a new software release, and there is a lot of work that went along with the deconstruction and reconstruction of the Tabernacle each and every time.

The classical medieval commentator Rashi offers a fascinating insight. You might even say that he alludes to the necessity for continuous integration as part of one's build process:

Moses said before the Holy One, "How is it possible for a human being to erect it [the Mishkan]?" God replied, "You work with your hand." Moses appeared to be erecting it, and it arose by itself. This is [the meaning of] what it says: "the Mishkan was set up" (Exod. 40:17). It was set up by itself.

The Torah describes the repeated constructions of the Tabernacle in the passive voice, "it was set up." It does not refer to who did the setup. Many people had responsibilities for various tasks in the breaking down and building of the Tabernacle, but the gargantuan task of actually erecting it fell on Moses' shoulders. How could he manage it? Well, he had some Divine intervention, but you could also say he built processes surrounding the construction that enabled his success. In other words, he built a continuous integration process that did a lot of the heavy lifting for him.

Do you have some form of a continuous integration process in your work? Are there automated tests that occur during deployment? Are there checks for visual integrity between builds? If you work in a developer adjacent role, what are your processes that help you deploy your work successfully and smoothly?

No one is exempt from needing tools to make their work possible, not even Moses.

LEVITICUS - VAYIKRA

1

VAYIKRA
WHEN SMALLER IS BIGGER

The laws of writing a Torah scroll are very exact. Halachah prescribes every detail of the process, from the ink and parchment used, to the shape of each letter.

While most letters are similar in size, some letters are written either larger or smaller than the surrounding letters.

The *Aleph* in the first word of this week's Torah portion, Vayikra (and he called [to Moses]), is one example. It is written smaller than the rest of the letters in the word.

The Ba'al Hatanya (the first Rebbe of the Chabad dynasty, 1745-1812) was asked about that small Aleph by his three-year-old grandson, Menachem Mendel (who would grow to be known by the name of his book, the *Tzemach Tzedek*, and would go on to become the third Chabad Rebbe, 1789-1866).

The Ba'al Hatanya replied:

> "There are two great people mentioned in the Torah. Adam was the first man, created by God's own hand, and his wisdom was greater than that of the angels. Adam was aware of his greatness,

and he let this greatness get to him. At the end, that was the cause of his downfall.

"Moses was also aware of his greatness. He was the only human being who spoke to God directly. Unlike Adam, in Moses, this awareness brought about a sense of humility. Moses felt that any greatness that he achieved was because God chose him for the task, and had God selected someone else and given his abilities to another person, that person would have reached the same heights, if not more.

"That is why," the Ba'al Hatanya continued, "when the Torah refers to Adam, it does so with a large Aleph (Chronicles I 1:1), but when it mentions Moses, it refers to him with a small Aleph."

Many times in life we are afraid to speak up, to stand up for ourselves or for what's right, because we've been taught to value humility. We feel that calling attention to ourselves and our achievements, or even just being conscious of them, would be considered vain and arrogant.

That mindset can sometimes be as harmful as the truly arrogant mindset we are trying to avoid.

The truth is that humility != (does not equal) low self-esteem or a lack of appreciation for one's achievements.

The true humility is when someone acknowledges their achievements, but is actively conscious of the privilege that got them there.

Using your position to lend that privilege to others and stand up for what's right is far from arrogance. It is the ultimate humility and showing *Hakarat Hatov* (gratitude) to all those who lent their privilege to us and helped us get to where we are today.

2

TZAV
ENCOUNTERING BURNOUT

Have you ever felt uninspired or simply just exhausted from work? Did the thing that you enjoyed doing so much become a source of dread? You used to enjoy writing code but now can barely open up your editor.

If you have experienced those feelings, you have certainly felt burnout. It is nothing to be ashamed of. Burnout happens to all of us at points in our lives. In many ways, it is a natural outgrowth of putting so much of your personal devotion and energy into an occupation.

This past week Jewish communities throughout the world celebrated the holiday of Purim. On Purim, we rejoice over the story of redemption from near annihilation. It is the story of Esther and Mordechai's bravery and heroism. It is the story of finding hope against all odds. It is also the story of encountering and overcoming burnout.

Nearly 1,000 years before Purim, the Jewish nation stood at the foot of Mount Sinai and experienced revelation. The Talmud (Shabbat 88a) records the opinion that the national act of acceptance of the Torah at that momentous occasion might not have been of total heart and mind. Indeed, a people who just recently experienced freedom from slavery and overwhelmed by the sheer miracle of the Exodus, might not be in the best position to secure a covenant for themselves and their descendants for all time.

Thus, the Talmud continues and teaches that a second acceptance of the Torah transpired, a thousand years later, at the time of Mordechai and Esther, during the holiday of Purim:

> "The Jews established and accepted"--they established [in the days of Achashveirosh] what they had already accepted [in the days of Moshe].
>
> (Talmud, Tractate Shabbat 88a)

Why was this necessary, and why specifically during Purim? We alluded to its necessity above when we mentioned the great debate in rabbinic literature. The nature of free will exhibited during the revelation at Sinai was qualitatively unique.

How free was their choice to accept it, really? As recently emancipated slaves, were they really in a position to make such an acceptance?

Thus, we find ourselves at a second pivotal decision point for the nation. They choose -- a thousand years later -- to truly, and with full free will, to accept the covenant offered at Sinai.

Yet, still, why during Purim and not any other moment? Rabbi Avraham Yehoshua Heschel of Apt, an 18th-century sage, offers the following idea:

> "... But even so, they accepted it in the days of Achashveirosh, which was a time of great pain and distance from God, and even so, they accepted it upon their souls, even though they were in a very lowly state, and committed to do all that was written in the holy Torah ..."

In other words, the Jewish people were not at a particularly high point in their national character during Purim. They had largely forgotten who they were as a nation. Moreover, at the exact same time, the Jewish nation was rebuilding its presence in the Land of Israel and returning from exile. But, these members of the community, the ones that were in Persia for the episode recorded in the Purim holiday, were the ones who chose not to go and rebuild their homeland and to not return from exile.

However, it was precisely at that moment, and with those individuals, that the powerful and tremendous moment of reacceptance of the Torah as free people had to occur. Rabbi Heschel suggests that it was exactly because they felt so disconnected, or burned out, that they had to go through the process of rediscovery and reaffirmation.

The Jews during the time of the Purim story in the land of Persia were distant from their collective call as a people and were burned out from their story. It had been a thousand years since Sinai, and they were tired. Nonetheless, they found a way to reclaim their passion and their inspiration and reaffirm what was central in their lives.

This is the challenge we face when we encounter burnout. How can we reclaim what used to motivate us and what used to drive us? The story of Purim can be seen as one people's attempt to encounter and overcome burnout. Let us find a way to use their success as an inspiration for us.

3

SHEMINI
TO ERR IS DIVINE

The other day we had a discussion at work over certain mistakes that were made, and what we as a team could learn from them. That got me thinking about how the Torah does "error handling."

None of us are perfect; we all fail sometimes. The question is what do we do with those failures.

We can sweep them under the rug and forget about them -- or at least try to, in the end, we all know they will come back to bite us -- however, that would make the whole experience a waste, a waste of time, effort, and a wasted opportunity.

Another option is to use these failures as an opportunity to learn from our mistakes and apply those lessons to help us, and others, avoid the same mistakes in the future.

This option transforms the failure. The failure is not seen anymore as a negative experience; instead, it is seen as the catalyst for growth, as the first step that allows the person to achieve the greatest heights. In fact, the Talmud says that "in the place where repentant people stand the most righteous people cannot stand." The righteous person who never experienced the struggle of failure can not achieve the heights attained by someone who struggled, failed, and came out on the other end a better person.

The same can be said of our work in tech. Failure is virtually guaranteed in our jobs; what we do with those failures is what defines us as IT professionals.

We can either dig our heads in the sand, pretend nothing happened, or if it's so bad that hiding it is impossible, assign the blame to someone else. The result of that will be fostering a toxic environment where, ironically, failures will keep happening with increasing frequency.

The better option would be to accept the failure, learn from it, and use it as an opportunity to grow as a developer and as a team.

What is the process of this transformation? How can we recast a negative into a positive?

One way is by applying the Torah's algorithm known as Teshuvah (usually translated as "repentance," though I prefer the more accurate translation of "return").

Teshuvah requires three steps:

1. Acknowledging your wrongdoing.

2. Verbal confession.

3. Resolution for the future.

Applying these three steps to our lives in tech, we can see that to truly learn from our mistakes we must:

1) Own up to the fact that we did, in fact, make a mistake.

2) Talk it over, discuss it as a team, analyze what processes and blind-spots allowed this mistake to occur, and

3) Make action items to implement the lessons learned so that we don't repeat them in the future.

Failures provide us with valuable opportunities. If we learn not to waste them, we will see that while to err is human, growing from our errors allows us to touch the divine.

4

TAZRIA
DEALING WITH UNHEALTHY
ENVIRONMENTS

Have you ever worked in an unhealthy work environment? Did you ever have a colleague that made you feel uncomfortable? Did you ever have a manager that denigrated you, insulted you, and put you down regularly? What did you do in that situation?

In this week's Torah portion of Tazria, we encounter a highly technical discussion of skin disease and treatment. It can feel out of place for what we might expect from a religious text. Yet, in the course of that conversation, we discover a fascinating idea.

While a person who has been diagnosed with a certain skin disease called tzara'at (which is hard to translate into English), that person must be isolated from the rest of the community. You could imagine one reason for that exclusion is containment of any possible contagious disease. However, an additional element, a more spiritual perspective, is added by the Talmud in Tractate Arachin 16b.

There, in the Talmud, the Sages offer the perspective that one reason the person is excluded is because they are suffering from a disease brought about by their exclusionary activity. In other words, they were afflicted with this skin disease precisely because they spoke slander and ill will of other people and brought about separation in society. As a result of this, they are compelled to experience communal separation. It is an exercise in empathy building for the offender, and also a chance for the community to heal.

In that idea lies a powerful notion for us nowadays. Sometimes there are moments that call for immediate reconciliation and coming together. There are times to work on making things better. And, at the same time, there are also moments that call for separation and removal. It is just the fact that sometimes an unhealthy work environment will always be unhealthy, and the only thing that can change is your place in it.

The Torah in this week's portion teaches us that it is okay to remove ourselves from negative situations at times. In the original conception, it was the offending person who was removed from the situation to experience the exclusion they brought about by their offensive words. Nowadays, since we can't always compel our colleagues or supervisors to become productive members of the workplace, we can exercise that removal for ourselves and take ourselves out of those negative situations.

5

METZORA
PROVIDING EMPATHETIC FEEDBACK

In last week's issue of Torah && Tech, we spoke about how we can learn from the laws of the Metzora that sometimes the best way to deal with a toxic situation is to remove the toxicity so that we can maintain a positive, healthy environment.

How do we put that in practice?

Suppose you're a senior member of your team, and it's time to provide feedback to the new developer. You're chomping at the bit. You've been bottling up opinions, and boy can't you wait to let them know.

This developer is terrible! Their code consistently breaks production, their variable names follow no convention known to man, and worst of all, they use tabs instead of spaces!

Of course, part of you feels guilty for chewing them out, but hey, "it's for their own good," says that little voice in your head; even the Torah says that sometimes you have to remove toxic influences from your life and your team.

There's an interesting law regarding the Metzora. When a person suspects that they may have *Tzara'at*, they must go to a Kohen, and only the Kohen can pronounce them pure or impure. Even if the Kohen does know the complex laws of *Tzara'at*, he must consult a scholar who will inspect the afflicted person and tell the Kohen to "say *Tameh*" or "say *Tahor*," and only once the Kohen makes his pronunciation would the person become impure.

The question begs itself; why the need for the Kohen? If it's the scholar who makes the determination, why can't he "cut out the middleman" and pronounce the person impure himself? Furthermore, there is no other state of impurity with a similar restriction, why is *Tzara'at* singled out as the only one that requires a Kohen to establish the state of impurity?

The answer lies in the severity of the *Tzara'at* impurity. *Tzara'at* is the only state of impurity in which the person was removed from the community. A pronouncement of *Tzara'at* was, in effect, a verdict of excommunication. Before handing down such a "*Psak*" (verdict), the Torah wanted to be sure there were no dark motives involved.

To that effect, the Torah requires that any pronunciation of *Tzara'at* would go through a Kohen. Kohanim are the essential lovers of Israel; they are descendants of Aharon, who was famous as the "lover of peace and pursuer of peace." The blessing the Kohanim make before the *Birchat Kohanim* says they are commanded to "bless the Jewish nation with love."

Only a Kohen, the epitome of unconditional love, could be trusted with the ultimate of all punishments. If the Kohen pronounced someone a Metzorah, we could be sure he explored all avenues to avoid the situation first, and we could be sure he would do everything in his power to resolve the situation and pronounce the person Tahor as soon as possible.

So should you avoid giving critical feedback? Of course not! Constructive, actionable feedback is how we all grow. But before you do, take a close look at your motivation.

Is the feedback coming from a place of empathy and a desire to improve, or are there darker forces at play?

If we make sure that our feedback is coming from a place of empathy, we will find that, not only will it not contribute to a toxic situation, but it will be accepted much easier and have the desired results more often.

6

ACHAREI MOT
WE CAN DO BETTER!

This week's Dvar Torah is dedicated in loving memory of Lori (Leah) Gilbert-Kaye ה״יד who was murdered in the shooting attack at the Chabad of Poway Shul. May her memory serve as a blessing.

Last week we were all shocked by the news that came in from Poway, CA. The fact that the shooting at the Chabad House happened during *Acharon Shel Pesach*, traditionally a joyous time for Jews worldwide, and only six months after the shooting in Pittsburgh makes it all the more incomprehensible.

At times like these, it's hard to come up with an appropriate and coherent response, yet I believe that we techies, particularly techies who hold Torah values dear, are perhaps better equipped to respond.

The Talmud (Bava Batra 10a) tells of a debate between the great Talmudic sage Rabbi Akiva and the Roman provincial governor in Israel at the time, Tineius Rufus. The following is a loose translation:

> Tineius Rufus the Evil asked Rabbi Akiva, "if your God loves the poor so much, why doesn't he provide them with sustenance himself?" Rabbi Akiva replied, "So that we may merit the world to come by helping them."

Tineius Rufus said, "to the contrary, helping the poor makes you guilty in God's eyes, let me give you an example: Suppose the emperor got angry at one of his servants and had him thrown in prison, he commanded all of his other servants not to feed him. Now suppose one servant defied the king's orders and fed the errant servant, wouldn't the king be angry at him?"

Rabbi Akiva replied, "I will give you a more accurate example: Suppose the emperor got angry, not at his servant, but at his son, and in his temper he had him thrown in prison, he commanded all of his servants not to feed him. Now suppose one servant defied the king's orders and fed the errant son, wouldn't the king send him gifts when he found out? Similarly, we are all God's children, as it is written, 'you are children of the Lord, your God' (Deuteronomy 14:1)."

This argument may sound arcane, but it's strangely relevant. Tineius Rufus was committing a well-known logical fallacy known as the Just-World Fallacy. We all have an innate need to feel that the world is just and fair, so if we see someone suffering, say they are poor for example, we try and justify it by saying that the person must be lazy, or that they lack ambition, all as a way of justifying the seeming unfairness of their poverty. When taken to the extreme, as Tineius Rufus did, this argument can be used to defend why we don't do more to help them out.

Rabbi Akiva replied that such a world-view is false. When we see imperfection in the world, we need to see it as a challenge, a place we can improve.

This idea is especially relevant to us in tech.

One of the first things I had to learn when I was learning to program was how not to be afraid of error messages.

One of my favorite quotes that helped me get through that mental block was from Steve Klabnik:

> "Programming is a movement from a broken state to a working state. That means you spend the majority of your time with things being broken. Hell, if it worked, you'd be done programming."

"Civilians" are afraid of error messages. It's this scary thing the screen barks at them, which may or may not be their fault, and usually something they can't fix without the help of IT. As programmers, we see lots of error messages, but for us, error messages are guides, they are opportunities, they show us where the code broke and how to fix it.

We thrive in broken environments. We view systems that aren't working as an opportunity to fix and improve.

While other people may see darkness in the world and shrug while saying, "what can you do?" or even "I guess that's the way it is..." We know better. When we see evil in the world, we see it as an annoying error message, and if there's one thing we know about error messages, it's that we can eventually get to the bottom of it. It might not be easy, it might not take a few minutes, but if we persist, there's no bug we can't eliminate.

Let us take the feelings of anger and outrage and use them as a motivator to make the world a better place.

7

KEDOSHIM
FINDING UNITY WITH OUR
DIFFERENCES

We live in a world of so much strife nowadays. It is hard to turn on the TV news or read the paper without encountering violence between people, whether it is a violence of words or violence of actions. We are ever-closer to each other thanks to advances in technology, but at the same time, ever more distant. Can we find some wholeness, some way of coming together?

This past week was my (Ben) first time commemorating and celebrating, respectively, *Yom HaZikaron* (Israeli Memorial Day) and *Yom HaAtzmaut* (Israeli Independence Day) as a citizen of Israel. I made *aliyah* (immigrated) to Israel last summer. After a lifetime of dreaming and planning and hoping, we made the plunge and moved to this special place in the hearts and imaginations of our people.

However, upon coming here, you learn something very quickly: This place, most of the time, is just like every other place. There is traffic that makes you frustrated. There are customer service agents that can make you want to pull out your hair. There is the daily routine that you find everywhere, routines of working and leisure times. So much of what we do here is what we do everywhere.

Yet, this week was different. This week we connected with what makes us unique.

All too often, when we search for diversity and inclusion, we are actually looking for uniformity. One person recently remarked to me that he would accommodate diversity around special needs, race, and gender because the person has no choice, but he will not respect cultural and religious diversity, because a person chooses to make themselves different.

In other words, the zeitgeist seems to be as long as your diversity is from a place that you cannot control, I will respect it, but if you choose (bracketing whatever that word means) to live according to your values, well then you forsake all claims to inclusion.

This, my friends, seems to be a root cause of so much intolerance and division in our world nowadays.

I say, let people embrace and discover what makes them unique! In fact, a large part of what makes us human is the diversity of ideas that are part of the fabric of the human story.

This past week I found myself working remotely in a busy and bustling coffee shop. There was music playing in the background. The sound of the espresso machines kept on being heard. Kids were sitting at a nearby table, laughing and joking. A woman next to me was focused on composing an email on her laptop. A couple was having a loving exchange a few tables over. This scene at this coffee shop could have been in any place, anywhere in the world.

Then, just as the clock hit 11 AM, an air raid siren was heard. Everyone stopped what they were doing, stood up, and became quiet. The kids stopped playing. The woman stopped working on her email. The couple stopped their exchange. The espresso machines went quiet, and the music was turned off. Outside, cars stopped, and the drivers got out of their vehicles. For two minutes, this scene not only played out in my coffee shop in my neighborhood but also in every city throughout the entire country.

We stood and remembered the thousands who died to establish and then to defend Israel in the past 71 years. We stood and remembered the Jewish people who have been killed throughout the world because they were Jewish, including in Pittsburgh and San Diego in recent months and days. This is our unique story. It is particularistic and deeply personal. It makes our nation separate, in the same way that other nations have their defining days and defining stories that make them unique.

The two minutes of stillness ended, and life resumed. The music turned back on. The woman returned to her email. The couple held hands again. The kids started laughing once more. Two minutes in the course of a day to be marked as different. Two minutes to remind us that we strive for unity, not uniformity. Two minutes to remember that we can find true inclusion and diversity through embracing our humanity, which is made up of our freedom to choose lives of personal meaning.

We are in a cultural moment where diversity work seems to end where ideas and choice begins. This past week in Israeli society, a week of deep introspection, of memory, and of uniqueness serves as a reminder to find common cause with others with our differences, not despite them. To strive for a loving embrace of all people, including the values they choose to live by, not shunning those values and asking them to hide them in the public sphere.

8

EMOR
I MENTOR THEREFORE I AM

ויאמר יהוה אל־משה אמר אל־הכהנים בני אהרן ואמרת אלהם

"And Hashem spoke to Moshe and told him: speak to the Kohanim, the children of Aharon, and you should say to them…"

The above verse, the opening verse of Parshat Emor, sounds a little awkward. Why does the Torah, famous for its conciseness, repeating itself by saying "speak… and you should say"? The Talmud routinely learns lessons from superfluous letters; what can the addition of two seemingly unnecessary words teach us?

Rashi, the famous 11th-century commentator on the Torah, quotes the Talmud (Yevamot 114a):

"'Speak… and you should say' [This double expression comes] to admonish the adult [Kohanim to be responsible] for the minors."

There's a lot of discussion in tech around what makes someone a "senior" engineer. Do they write better code? Can they debug in their sleep? Are they able to pick up new technologies faster?

The best definition I've seen can be summed up in this tweet:

"The job of the senior developer is not to write more or better code than the junior or to review everything the junior did.

73

Their job is to help the juniors become more effective. To teach, coach, and mentor them. To help them become seniors."

— David Tanzer (@dtanzer) October 23, 2018

The Torah teaches us that it's not enough for us to learn on our own, instead we have a responsibility and an obligation to pass our knowledge on to others.

During the weeks between the holidays of Pesach and Shavuot, many have the custom to learn a weekly chapter from the *Pirkei Avot* (Ethics Of Our Fathers - a tractate of the Mishnah that deals primarily with ethical teachings). The very first paragraph of the *Pirkei Avot* encourages us to "raise many students." The Mishnah doesn't qualify its statement, implying that as long as there are students you can reach, you have not fulfilled your obligation to teach "many students."

Of course, you don't have to wait until you are "senior" before you start teaching and mentoring others. The Lubavitcher Rebbe OBM was fond of saying that "if you only know Aleph, don't wait until you learned the whole Torah, find someone who doesn't know Aleph and teach them!" In fact, there is no better way to solidify your knowledge than by turning around and teaching those behind you. Nevertheless, the more we grow in knowledge and experience, the more it is imperative for us to share that with those coming behind us.

And just in case there's that little voice in the back complaining about the waste of valuable senior engineering hours that can be better used elsewhere, let me conclude with one of my favorite Talmudic sayings:

"Rabbi Chanina said: I have learned a lot from my teachers, and I learned even more from my colleagues, but from my students, I have learned the most."

9

BEHAR
A LITTLE HELP CAN GO A LONG
WAY!

This week's Torah portion discusses, among other things, how to help others who've come across hard times.

וכי־ימוך אחיך ומטה ידו עמך והחזקת בו גר ותושב וחי עמך

"If your brother becomes destitute and his hand falters beside you, you shall support him..."

(Vayikra 25:35)

At this point, the Torah doesn't yet go into specifics regarding how you should support the person whose hand faltered, but Rashi fills in:

"You shall support him: Do not allow him to fall down and collapse altogether, in which case it would be difficult to pick him up again [from his dire poverty]. Rather, "support him" while his hand is still faltering [for then it is easier to help him out of his trouble]. To what can this be compared? To a load on a donkey; while it is still on the donkey, one person can grasp it and hold it in place. Once it falls to the ground, however, [even] five people cannot pick it up."

The lesson is a simple one; help keep people on their feet, it's a lot easier – and cheaper – than lifting them once they fall, though as a Torah observant techie, I think there are a few more lessons we can learn, relevant to the field.

The first is concerning our code. How many times do we come across old "legacy" code, and it's just a mess! Spaghetti code all over the place, nothing is documented, and every time you touch a line of code, something seemingly unrelated five files away seems to break.

How did it get that way? No one consciously decided to sit and write terrible code. It was the result of many small decisions. Maybe a tight deadline meant documentation was left "for next week." Perhaps a rushed manager made the decision that writing tests was a waste of time better spent on churning out features. Whatever the case was, small decisions were made, and the compound interest on the technical debt grew from there.

How many times were we working on a small project and cut corners because we figured it was such a small codebase, who cares about best-practices? Only to find that by the time those mistakes came to bite us in the back, fixing them would require a complete rewrite?

Spending just a few extra minutes writing a test, refactoring a function, and writing a README will save us hours of grief down the line!

The second lesson is in people skills, (and may not be limited to tech!).

Many times we find ourselves in a situation where a member of the team acts in a toxic way and is bringing the rest of the team down. At that point, the situation is very hard to deal with, and tough decisions often need to be made. But most times, those situations don't develop overnight, it usually starts with a small comment here, a snide remark there, and people don't want to say anything for fear of "overreacting."

Most people don't start off wanting to be toxic team-members. %99 of the time pulling someone to the side and mentioning that the joke they just made was a tad insensitive or that their comment during the meeting could have offended that intern can serve as the gentle nudge needed to make sure more significant problems don't manifest down the line.

To conclude with a something Rabbi Yosef Yitzchak of Lubavitch was fond of saying: when you see someone lost in the middle of a forest, they never got there all at once, it always starts with taking one step off the straight path.

Let's take care of our problems when they are small before we find ourselves supporting the whole load with the donkey too!

NUMBERS - BAMIDBAR

1

BAMIDBAR
A MORE COMPASSIONATE
WORKPLACE

Should we make a special effort to welcome junior professionals into our teams and into our companies? Wouldn't it make more sense to only hire senior people who bring with them years of experience and knowledge? There would be no need for training and for mentorship. The new team member could get started being fully productive right away. Who needs junior developers or junior designers or junior product managers?

Let's put aside for a moment the inherent challenge that if we only hire senior folks, then we will never grow the pool of senior-level people. I want to take a step back and examine the intrinsic value of mentorship through the lens of this week's Torah portion. Indeed, mentorship is not only valuable for the one being mentored, but it is equally valuable to the one mentoring.

(A quick note: This is the time of year when the weekly readings are out of sync between Israel and the Diaspora. This week in Israel, the portion being read is Bamidbar, while in the Diaspora, it is Bechukotai.)

The Torah in Bamidbar 3:1 states:

ואלה תולדת אהרן ומשה ביום דבר יהוה את־משה בהר סיני

"These are the descendants of Aaron and Moses on the day that God spoke to Moses at Mount Sinai."

The Torah then continues to list only the descendants of Aaron. The great medieval commentator Rashi notices this discrepancy and offers a brilliant insight:

"Yet only the sons of Aaron are mentioned. However, they are considered descendants of Moses because he taught them Torah. This teaches us that whoever teaches Torah to the child of his fellow, Torah regards it as if he had begotten him."

In other words, the mentorship relationship has the potential to be elevated to something resembling a familial bond. It is not only that the mentee gains valuable knowledge and wisdom, but through the act of mentorship, both members of the relationship -- mentor and mentee -- become transformed through it.

All too often, mentorship of new colleagues can be seen as a chore imposed on more senior members of the team. At its best, it is perceived as an enjoyable part of the job, but just a task nonetheless, and at its worst, it is viewed with contempt and anger.

Yet, what if we saw it for what it truly can be? Mentorship has the potential to create closer, more well-functioning, and holistic teams. It has the potential to introduce our human selves into our working relationships, which can catalyze more empathy, more compassion, and more understanding. It is not just a necessary aspect of an industry that needs to constantly grow its talent pool, but an intrinsic aspect of cultivating more people-centered workplaces.

2

NASSO
VALUE EVERY CONTRIBUTION

What makes for a valuable contribution to the projects we work on? This is a hotly debated issue in the developer community. Do grammar fixes count as open source contributions? Does a contribution need to be only code for it to be deemed valuable?

In Parshat Nasso, we find ourselves at the completion of the construction of the Mishkan, the Tabernacle. This was one of the greatest endeavors of the Jewish people while they traversed through the wilderness. This massive construction project, which needed to be constructed and deconstructed each time they moved, required the active participation of all the people. It could be seen, in fact, as a nationwide open-source initiative.

At the conclusion of its construction and during its inauguration, the heads of the tribes offer gifts. Each tribe, through its head, offers a gift on a separate day. As we are reading the description of the gifts, we discover that each gift was identical! Why so?

There are many beautiful and profound teachings associated with this question. However, let's focus on one idea:

Each person's contribution is valuable.

It's a simple idea with important ramifications. Whether the contribution was the same as the person before or was a "small" grammatical fix, they are all valuable. A contribution of time, resources, and energy is an expression of a personal commitment to the project and to the ideas that it represents.

The heads of the tribes could each have offered their own unique and elaborate gift. After all, they were leaders of their communities. Yet, they chose to send a powerful message that reverberates down to this day. Let us not ignore the gifts of any person. Let us appreciate all that people bring with them in service to the projects we care about, the communities we invest ourselves in, and the world we inhabit.

3

BEHAALOTECHA
DO YOUR PART

As developers, we often find ourselves dealing with problems larger than ourselves. Sometimes it seems like we just can't do it all on our own.

There are usually two ways people go about coping with such situations.

There are those who see themselves as martyrs. They will put in long hours, spend weekends in the office living off of pizza and Red Bull trying to get it done, sacrifice their health, and their social life. Then, at the end of it all, when things inevitably fall through (because the problem is, after all, larger than any one person), they will take it all very personally, blaming themselves for the failure and feeling completely worthless.

Others will take the opposite approach. They are the realists, the pragmatic ones. When they see the problem is too large, they will throw up their hands and go home knowing it's not their problem.

Then there's the third approach.

There's a beautiful custom that many people have during the six weeks between Passover and Shavuot to learn a chapter from the Pirkei Avot (a tractate of the Mishnah that collects ethical teachings and maxims) every week, finishing the six-chapter tractate in time for Shavuot.

Some communities continue this custom throughout the summer and, following that schedule, will be learning the second chapter this week.

The last Mishnah (teaching) of the second chapter contains a powerful thought that I find myself going back to often (Avot 2:16):

לא עליך המלאכה לגמור ולא אתה בן חורין לבטל ממנה

[Rabbi Tarfon] would say: It is not incumbent upon you to finish the task, but neither are you free to absolve yourself from it.

Rabbi Tarfon is speaking of a much larger task than any we might come across in our work, the task of transforming the world into a Godly home. Nevertheless, the lesson he taught can be applied to our lives as well.

None of us lives in isolation; others will help us and continue our work when we're finished.

The work may not be ours to finish, but by not absolving ourselves from doing our part, we get to become part of something much larger than ourselves.

4

SHLACH
CHALLAH IS MORE THAN BREAD

Last week we discussed the idea that one should set realistic expectations for their work. You don't always need to be the hero and shouldn't set unrealistic goals for yourself all the time. However, It is also okay to rise above once in a while and push yourself -- all within balance.

This week, Parshat Shlach presents us with another aspect of work-life balance, and this time it comes in the form of bread.

The Torah instructs the people to begin to set aside a portion of every bread they make as a gift to God. This is known in Hebrew as hafrashat challah, meaning the separation of the challah. (Quick fact: The word "challah" is actually referring to the gift portion of the bread, not the braided bread we eat on Shabbat and holidays.) Bread represents, in society, a major act of work. It is a staple food for cultures across the world. Bread, as we know, does not grow from the ground, but rather is the culmination of work and effort to create it from the raw products that go into it.

This work and this effort are not meant to be all-consuming (pun intended) of our attention and focus. Rather, the Torah instructs us to apportion a part of the physical and emotional labor into a higher purpose.

Can we find a higher calling for a portion of our work and the focus of our days? Is there some "challah" that we can take from the hours we expend every single day on our projects and our tasks? Last week we discussed the balance of doing your part versus self-care, and this week, we raise the question of finding meaning and dedicating a portion of your work to it.

As you enjoy your "challah" bread this Shabbat, find some time to think about the ways you can take challah -- dedicate a part of your daily efforts -- to something greater.

5

KORACH
HOLY TECH!

Note: This coming Shabbat, the 3rd of Tammuz in the Hebrew calendar, marks the 25th anniversary of the passing of the Lubavitcher Rebbe of blessed memory. While the Rebbe is no longer with us physically, his influence is still very much felt throughout the Jewish world and beyond. We dedicate this issue of Torah & Tech to his memory and inspiration.

In January of 1960, a controversy broke out. A Chabad Chassid by the name of Rabbi Yosef Wineberg started a weekly series of Torah lectures on the radio. Many in the orthodox world at the time felt it was inappropriate to spread such holy Torah thoughts over a medium commonly used for secular entertainment that often stands at odds with traditional Torah values.

The controversy reached the Lubavitcher Rebbe, Rabbi Menachem Mendel Schneerson OBM, who couldn't disagree more.

In a public gathering, called a Farbrengen, the Rebbe clarified his position on the use of technological advances:

> "Our sages have said: 'All that God created, he created for His honor.' This also applies to all the scientific discoveries of recent years—their purpose is to add honor to God by using them for holiness, Torah and mitzvot...

"There is a particular advantage to using radio to teach Torah. Even if a person is not sufficiently motivated to go and attend a class in person, or even if he only turned on the radio to hear something else—the words of Torah reach him.

"Furthermore: also in a place where there is no human being to hear these words or not even a radio receiver to make them audible, the words themselves permeate the place, achieving the end of 'spreading the wellsprings of divine wisdom to the outside.'"

This attitude was a trademark of the Rebbe's leadership over the decades, and his followers embraced it wholeheartedly. At a time when a large part of orthodox Jewry viewed modern technological advances with suspicion, if not open hostility, Chabad Chassidim embraced them as a tool for spreading Godliness through the world.

It didn't stop at radio either. The Rebbe's followers went on to pioneer the use of live television, satellite hook-ups, and, eventually, the internet to try and make the world a more Godly place.

In fact, Chabad presence on the internet goes back to 1988 and predates the creation of the World Wide Web itself! Chabad.org, Chabad's flagship website was founded in 1993, before Google, Yahoo, or Wikipedia were online!

It is this attitude, that everything God created can, and must, be used for good that continues to inspire countless people worldwide to use all of the tools at their disposal to increase goodness and kindness in the world.

6

CHUKAT
BRUTE FORCE A SOLUTION?

In the work of technology, we can often find ourselves frustrated. We can be frustrated when the thing doesn't work the way we want it to or doesn't work at all. We can be frustrated that things are just not going our way. "I am following the documentation exactly; why isn't it working?!?"

If you have ever experienced that moment, you can relate to the desire to just come to a solution, any solution. Regardless of how forced your workaround may be, if it gets the job done, at least you can say you are finished.

Moses, too, experienced this sort of frustration. In this week's portion of Chukat, we find ourselves in the midst of the desert and mourning the loss of Miriam, his older sister. It is at that very moment that the people begin to quarrel. They start arguing with each other, and with their leadership, with Moses and Aaron. One of their leaders from the very beginning has died, they are out of any water, and they are tired, thirsty, and anxious.

> "Why have you brought the congregation of the Lord to this desert so that we and our livestock should die there?" (Numbers 20:4)

These are the words of angry and desperate people.

God instructs Moses to take care of their most immediate and pressing need: the lack of water. He is to do so through a miraculous speaking to a rock, wherein water would pour forth. Moses, though, is also frustrated, and so he decides rather to brute force the problem, and he hits the rock. The result is the same, but the path to get there is markedly different.

A people embroiled in conflict were to witness the healing and restorative powers of proper speech embodied in this miracle of water from a rock. Rather, they saw their problem resolved through more aggression.

The consequences for Moses are severe. He is not to enter into Israel with the rest of the nation, but rather must remain behind. The people are about to enter into a new era of existence, where the demands of national life in their own home could brew many potential disagreements, and they require leadership that can handle that pressure, and not brute force it.

A lot of the time, it is not just about getting to a solution, but the way we got there is equally important. Taking a step back, assessing the situation, and acting calm and cool under pressure, can let us achieve better thought out solutions that achieve long term success.

7

BALAK
BILAAM THE HACKER

This week's Torah Portion contains the iconic story of Balak, the king of Moab, who hired the sorcerer Bilaam to curse the Jews. To their consternation, God took control of Bilaam's tongue and switched all of the curses he intended to utter into blessings.

The story usually ends there and is often brought as an example of God's undying love to the Jewish people, but the truth is that there's a continuation. It turns out that after Bilaam's plan to curse the Jewish people failed, he advised King Balak to entice the Jewish People to sin. Balak took Bilaam's advice; he sent Moabite and Midianite women to the Jewish camp to seduce the Jewish people and convince them to serve the Midianite god Pe'or. The plan worked, and a great plague broke out in the Jewish camp, killing thousands.

Suddenly the story doesn't seem so happy anymore! So the Jews won one battle, but at the end of the day, it sure seems like it was Bilaam who had the last laugh?

To understand why Bilaam lost this game, you have to understand what he was trying to accomplish.

You see? Bilaam was a master hacker.

Back in the day, hackers used to sit at their terminals, and do their best to break into systems. Companies didn't want people breaking into their systems, so they invested plenty of resources to make their systems more secure. So a mini arms race started where companies would build better and better defenses around their systems, and hackers would work harder and harder to find vulnerabilities allowing them to break in.

Eventually, the hackers gave up. Security just got too good to make their efforts worthwhile. You still hear of the occasional breach here and there, but the majority of hacking moved to what's known as "social engineering hacks."

In simple terms, the hackers realized that instead of breaking into the well-protected servers of a company like Google for example (a nearly impossible task), it's a lot easier to send a mass email to millions of email addresses at a time offering free vacations to Hawaii to anyone who would just fill out the form with their username, password, and social security card number. Out of the millions, there are bound to be a few people who won't think twice, they'll fill out the (fake) form, and before you can say Aloha, their entire contact list is on a remote server in Nigeria. In effect, the hackers discovered that instead of breaking into secured systems to retrieve people's information, it's a lot easier to get people to hand over that information on their own.

That, in short, is the story of Bilaam.

Bilaam knew there was a "secure connection" between God and the Jewish People. Nevertheless, like all good hackers, Bilaam tried finding vulnerabilities that would allow him to inject some malicious code.

Bilaam tried exposing the Jewish People's sins and showing them in an unflattering light, all in an attempt to inject some virus or malware that would get in between the connection between God and the Jewish People.

In that, Bilaam failed miserably. He discovered, and proved to the world in the process, that the connection between God and the Jewish People is secure and unbreakable.

Naturally, this was disappointing to Bilaam and Balak, and in desperation, they went for the next best thing, they "social engineered" a hack where they would get the Jewish People themselves to sever the connection between themselves and God.

This approach was successful, but only to an extent. Getting Jews to sever the connection between themselves and God is as useful as gaining access to someone's account by phishing their password; it only works until the victim realizes what happened and changes their password.

Similarly, the disconnect that occurs when a Jew leaves God is temporary and only lasts until the person realizes what they had done and decides to return; as soon as a Jew does Teshuva and returns to God the connection is reestablished, and much more securely than before.

Bilaam and Balak may have won one particular battle, but the cyberwar has already been decided in favor of the Jewish People.

8

PINCHAS
ADVOCATE FOR YOURSELF

How many times have you felt uncomfortable standing up for yourself and your own needs? How many times have you felt uncomfortable saying, "I did that!" when someone asks who did that great project at work?

It can be uncomfortable to advocate for yourself. Yet, if I am not for myself, who will be for me? (The inverse is equally true: If I am only for myself, what am I?)

In this week's Torah portion, the daughters of Tzelafchad advocate for themselves and their families. Tzelafchad passes away, and he has no sons. His portion in the Land of Israel would be lost to his family according to the law as it was understood at that moment. However, his daughters take that uncomfortable step and say something. They speak up. They bring their claim to Moses, who does not know what to do, so he brings it to God. In so doing, the law is revealed to be in actuality in favor of the daughters, and their larger families.

If they had not advocated for themselves, who knows how long it would have taken for it to be revealed that women can inherit the property of their families? How many families would have gone through the pain of losing their inheritance before it would have changed?

In advocating for themselves, they ended up improving the lot of the entire nation.

The next time you feel uncomfortable about standing up and advocating for yourself, remember the daughters of Tzelafchad. You might end up making things better, not just for yourself but for others as well.

9

MATOT-MASEI
SEE SOMETHING? SAY SOMETHING!

As Jews, we know that the only way to ensure our future is by looking back to our history, both the happy parts and the sad. This time of year, we find ourselves in the mourning period between the 17th of Tamuz (the day the walls of Jerusalem were breached by the Romans) and the 9th of Av (the day the Temple was destroyed).

The Talmud (Gittin 55b) relates that the destruction of Jerusalem, like many of our most annoying bugs, was caused by a typo.

One of the wealthy citizens of Jerusalem was making a party and wanted to invite his friend, Kamtza. His servant accidentally invited a different man by the name of Bar Kamtza, a sworn enemy of the wealthy host.

When the wealthy man saw his enemy at the party, he was furious and demanded that Bar Kamtza leave immediately. In an effort to save face, Bar Kamtza offered to reimburse the host for the cost of the entire party, but the host wouldn't hear of it and had him thrown out in shame.

Bar Kamtza was humiliated! "All of the sages and scholars of Jerusalem were sitting there, and not one of them said a word? I will show them yet!"

Bar Kamtza traveled to Rome and told the emperor that the Jewish people were planning a revolt. The emperor sent his general Vespasian to investigate, and the rest is history.

Reading the story drives home an important point. What hurt Bar Kamtza most, even more than being thrown out of the party of the year, was the fact that no one stood up for him.

Many times we might see injustices taking place. It might be someone getting blamed for something they didn't do, or a marginalized team member not getting credit for their work (again), not to mention cases of deliberate harassment.

Speaking up in such situations is hard. It can make things awkward; it could pull us into a conflict we have no interest in being part of. At such times we must think of the cost of not speaking up! When we fail to speak up in such situations, we take an active role in making the victim feel isolated and alone.

When the cost of action seems high, we often default to inaction; What we have to realize is that the cost of inaction is usually higher.

Let us not repeat the mistakes of history.

By replacing the "baseless hatred" that destroyed the Temple with "baseless love," we will merit having the Temple rebuilt speedily in our days!

DEUTERONOMY - DEVARIM

1

DEVARIM
YES, YOU CAN... BUT, FIRST IT'S OKAY TO CRY

Shabbat this week falls on the most tragic day of the year on the Jewish calendar. The 9th day of the month of Av is the day in which Jews worldwide gather and mourn over all the tragedies that befell our people from the destruction of the First Temple in 586 BCE to the destruction of the Second Temple in 70 CE and on through the Crusades, the Inquisition, forced dispersions, pogroms and massacres, and the Holocaust. We mourn traditionally through abstaining from food and drink, avoiding the pleasantries of greeting each other, wearing nice leather shoes, and even sitting on comfortable chairs. Because the 9th of Av is on Shabbat, the observances are delayed until the end of Shabbat and conclude the following day after nightfall.

We do this every year because we know a fundamental truth. You cannot rebuild if you do not mourn. Yes, you can, but first, you can cry.

The Talmud teaches (Taanit 30b) that all those who mourn for Jerusalem will merit to see its rebirth. Life is full of mountains and valleys, of great joy and great despair. We can attempt to construct bridges that bypass the valleys of despair and allow us to coast from one joy to the next, but perhaps we will miss out on something intrinsic as a result of doing so.

This 9th of Av may we find the time, the courage, and the resilience to cry so that one day we can also laugh.

2

VA'ETCHANAN
BOUNCE BACK!

לא היו ימים טובים לישראל כחמשה עשר באב וכיום הכיפורים

"There were no greater holidays in Israel than the fifteenth of Av and Yom Kippur."

(Talmud Ta'anit 26b)

The Jewish calendar is full of ups and downs.

Sunday, Tisha Be'Av, we fasted to commemorate the destruction of the Temple in Jerusalem and the start of the Jewish diaspora. On the other hand, Friday, the 15th of Av, was declared in the Talmud to be on a level with Yom Kippur, the holiest day of the year!

As Jews, we're used to experiencing setbacks, but the Torah teaches us that every setback is an opportunity for growth.

I was reminded of this idea by a conversation I had this week. A friend of mine is taking some time off to focus on learning some new skills that will advance her career. During this time, she will be contributing less and was worried that perhaps this was a selfish act.

This got us into a discussion about how sometimes we need to take one step backward to get us ten steps forward.

Chassidic sources speak of a Yeridah Tzorech Aliyah (a descent for the purpose of a higher ascent).

Ever noticed how an athlete might take a few steps back before the run-up to a long-jump, or the way you bend your knees before jumping high? Similarly, sometimes, we need to take a step back if we want to propel ourselves forward.

The same applies to our work. Recently our team decided to take two weeks in which we did not work on new features; instead, we focused on cleaning up our working environment to bring everything up to date.

At first glance, it may have seemed like our work slowed down (or even stopped), but the gains in productivity we saw over the next few weeks more than made up for the lost week or two.

To quote Nelson Mandela: The greatest glory in living lies not in never falling, but in rising every time we fall.

3

EIKEV
YOU ARE MORE THAN YOUR JOB

You are more than your job. Your worth as a person is greater than the position you hold, and the work you do for a living.

Those of us who work in tech can find ourselves constantly enthralled with our work. Many of us get to work with new technologies, solve fascinating problems, and exercise our minds regularly. It can be exciting. We may end up wearing our company t-shirts, putting our company stickers on our laptop, and in general, showing full pride in the place we work. But, what happens when a job ends?

A job may end of our own will, or it may end against our will. However, regardless of how it ends, what do we do with our value and worth of self that is wrapped up in the job?

In this week's Torah portion of Eikev, Moses instructs the people that: "humans do not live by bread alone." (Deut. 8:3) We exist for more than the work we do on a daily basis. Rabbi Joseph ben Isaac Shor, a 12th-century French exegete, commenting on the verse states that this is coming to teach us not to place our sense of security on "our bread," meaning our livelihoods.

Our task is to find that greater sense of purpose that transcends the work we do for a living. This Shabbat, find some time to think about what your values are, what your priorities are, and what you truly hold dear in your life. Take some time to connect to those and lift them up as points of self-value that last beyond any one source of work.

4

RE'EH
ELUL - OUR ANNUAL RETRO

One of the core practices introduced by the Agile movement was the retrospective or "retro" for short.

At set times -- the end of the week, the end of an iteration, after finishing a project, etc. -- the team gets together to discuss how things went.

There are probably as many ways to hold a retro as there are teams that are holding them. What they all have in common, though, is that they are a time to discuss what went wrong, what went right, and how the team can improve in the future.

Having regular retros is an integral part of having a healthy team that keeps growing and learning from its mistakes.

In our spiritual life as well, if we don't set aside times to stop and take stock, we will find ourselves deteriorating quickly.

Every once in a while, we need to take a break, look back, see what we could have done better, and make a plan to improve.

This coming Shabbat, we begin the month of Elul, the last month of the year, and the month where we prepare for the High Holidays.

Chassidic sources called the month of Elul "The Month Of Accounting."

Just as our retros keep our teams healthy and ensure that we continue to grow, taking advantage of the month of Elul to introspect on the year and make resolutions on how we plan to improve next year will help us grow as people and as Jews.

5

KI TEITZEI
AM I MY CODE'S KEEPER?

To what extent are we responsible for the consequences of our work? Can we assume our users will use our products in the manner that we intended?

If I create an app to help people keep in touch with their friends, is it my responsibility to put in place proper safeguards to ensure it isn't used to stalk and harass people?

In this week's Torah portion we read:

כי תבנה בית חדש ועשית מעקה לגגך ולא־תשים דמים בביתך כי־יפל הנפל ממנו

"When you build a new house, you shall make a guard rail for your roof, so that you shall not cause blood [to be spilled] in your home if the one who falls should fall from it [the roof]." (Devarim 22:8)

This verse is the springboard for many pages of Talmudic discussion about the responsibility we have to ensure that others don't hurt themselves misusing our property.

But while we can all agree that protecting innocent victims is important, other cases are not so clear-cut.

What about malicious actors? The people who will use our product for nefarious purposes despite our well-meaning Code of Conduct? To what extent are we responsible for them?

After all, the person using our platform to harass others will probably be harassing them whether we allow it or not. If we block them on our platform, they will find another platform and continue their abusive behavior from there. Is it really our responsibility to protect people from themselves?

A deeper look into the verse above holds the answer.

There's an apparent tautology here; the verse says we should put a fence around the roof of our home lest "the one who falls should fall from it." "The one who falls" seems redundant; if the person fell off the roof, then obviously, they are "the one who falls."

Rashi explains: "that person may have deserved to fall; nevertheless, you should not be the one to bring about his death."

Even if the person getting hurt is "one who falls,"; they'll fall either way. If they don't fall off my roof, they'll find another roof to fall off of. It doesn't matter. It's your responsibility to ensure that your roof is safe and is not the cause of hurt and pain.

6

KI TAVO
THINK (ALSO) WITH YOUR HEART

Why was so much of the Torah written as stories?

It is a basic question. A question so fundamental, we may never have stopped to consider it before. We take for granted the books that line our shelves, and the inheritance passed down to us through the generations. Yet, if we pause and consider this most basic of questions, it raises a truly important notion for us.

I (Ben) recently sat down with an engineer who is interested in starting to write blog posts on his work. He has a lot of good ideas and lessons he has gleaned from his career, and he wants to share them with a broader audience. These will be highly technical blog posts covering topics such as DevOps, code coverage, security auditing, and more. What was my overarching advice to him? Don't forget the narrative you want your reader to go on. There must be a beginning, a middle, and an end. There must be a compelling story that brings the reader along in a blog post, even and especially technical ones.

However, the reason the Torah is written in story form is not just to make it compelling and provide a narrative journey. It's much more than that.

In this week's Parsha of Ki Tavo, Moses continues in his retelling of the people's physical and spiritual journeys through the wilderness. In doing so, he says:

לא־נתן יהוה לכם לב לדעת ועינים לראות ואזנים לשמע עד היום הזה

"Yet to this day God has not given you the heart to understand or eyes to see or ears to hear."

What does it mean to not possess a "heart to understand"? Indeed, many translations use the word "mind" in place of "heart" to make the verse more comprehensible. Nonetheless, the Hebrew is "לב - heart" not "mind." How do you understand with your heart?

Rabbi Levi Yitzhak of Berdichev, the great Hasidic master, commenting on this, says the Torah is not only here to teach us ideas. It is not only an intellectual pursuit but also an ethical and spiritual one. We learn values from the Torah, not just concepts.

Ethics and values are best conveyed through the usage of stories. This is why the Torah is composed as primarily a book of stories. We learn not just what to do and what not to do, but we learn how to be ethical people with values.

7

NITZAVIM
FEEDBACK? IT'S ALL IN HOW YOU EXPRESS IT

It is that time of year again. We are in the week between Rosh Ha-shanah and Yom Kippur, the two holiest days of the Jewish calendar. This time is called the Aseret Yemei Teshuva, the Ten Days of Repentance. It is a time to look back on our previous year and commit to lead a better one in the coming year.

In other words, it is about feedback. Feedback about ourselves and to ourselves.

Who likes to receive feedback? Well, maybe if we are pressed hard to remember the one or two people who gave good feedback, we could say we like to receive it once in a while. But overall, feedback is hard! This is true whether we are receiving it from ourselves or others.

The Talmud in discussing the topic of feedback offers important advice, especially during this season of repentance, and internal feedback:

> Rabbi Yehoshua ben Levi said: As the verse said: "And you shall know that your tent is in peace; and you shall visit your habitation, and shall not sin" (Job 5:24). Rabba bar Rav Huna said: Although the Sages said that there are three things a person should, indeed

is required to, say in their home on Shabbat eve at nightfall, one must say them calmly so that the members of their household will accept them from them.

(Shabbat 34a)

You want your feedback to be received? Try saying it with a kind smile, and with gentleness. This goes for whether you are offering it to yourself or to your colleague.

This Yom Kippur, as many of us will be in synagogue attempting to go through the hard work of introspection and commitment to achieve better in the coming year, may we remember the lesson from the Talmud. Speak to yourself with kindness, and you may just find you actually listen.

8

HAAZINU
YOU BELONG!

"CSS isn't a real language."

"If you don't have a Computer Science degree, can you call yourself an engineer?"

"Front-end development is so much easier than back-end."

"Designers aren't actually technical."

If you've been in tech long enough, chances are you've probably come across some variant of one the above statements before. What they all have in common is that they are exclusionary, gate-keeping, and said with the arrogant attitude of someone who thinks that "I'm the 'real' engineer, and anyone who hasn't had the same experiences I've had and taken the same path as me is clearly an impostor."

Not only does the attitude above reflect poorly on the person who has it, but it's also one of the main aspects holding people from underrepresented demographics from joining tech.

The Four Species; an exercise in inclusion.

Next week observant Jews all over the world will participate in a timeless Mitzvah. They will gather together a branch from a date palm, a myrtle tree, a willow bush, and a fruit of a citron tree. They will take this interesting bundle, make a blessing, and wave it in all six directions (Vayikra 23:40).

Why these particular four plants?

There is a Midrash that explains:

טעם ויש בו ריח, כך ישראל פְּרִי עֵץ הָדָר – אלו ישראל, מה אתרוג זה יש בו
יש בהם בני אדם שיש בהם תורה ויש בהם מעשים טובים... מה התמרה הזו
יש בו טעם ואין בו ריח, כך הם ישראל יש בהם שיש בהם תורה ואין בהם
מעשים טובים... מה הדס יש בו ריח ואין בו טעם, כך ישראל יש בהם שיש
בהם מעשים טובים ואין בהם תורה... מה ערבה זו אין בה טעם ואין בה ריח,
כך הם ישראל יש בהם בני אדם שאין בהם לא תורה ולא מעשים טובים...
אמר הקב"ה יוקשרו כולם אגודה אחת והן מכפרין אלו על אלו.

"A fruit of a citrus tree" - these [species] refer to the Jewish people. Just as the Etrog (citron) has a pleasant taste and a pleasant smell, so too with the Jewish people, some have Torah and good deeds. And just as the date palm has a good taste but no smell, so too with the Jewish people some have Torah but no good deeds. And just as the Hadas (myrtle) has a pleasant smell but no taste, so too with the Jewish people some have good deeds but no Torah. And just as the Aravah (willow) has neither a pleasant taste or smell, so too with the Jewish people, some have no Torah or good deeds... Said the Holy One Blessed Be He let them be tied in one bundle so they can complement and atone for each other.

The Midrash makes a compelling point. Someone might decide that they want to put together a group of "10x Jews"; instead of the proscribed four species, they'll put together four Etrogim, surely such a bundle would be superior in every way?

The Torah tells us no! Every person is special; everyone is unique. We all have something to add!

So next time you feel the need to question someone's "tech cred," remember, even without the Aravah (the willow without taste or smell), the entire Mitzvah is invalid!

HOLIDAYS

1

PASSOVER
MATZAH AND KEEPING IT SIMPLE

It is Passover! This holiday is one of the calendrical highlights of the Jewish holiday cycle. It is the story of our liberation and the story that defines us as a nation. We first are referred to as a "nation" in the context of the Exodus from Egypt. I (Ben) am currently sitting in a coffee shop in the mall in my city in central Israel writing this. The mall is crazed with people buying items for their family Seder. The food court is divided between the restaurants that still have bread, and will thus be closed for Passover and the restaurants that are temporarily closed in preparation to be open during Passover. Every person in the country is wishing everyone else a "Chag Sameach -- Happy Holiday" after every encounter.

There are so many ideas, so many valuable takeaways and lessons to be gleaned from Passover. Let us just focus on one area for our message today. Let's talk about simplicity.

As a developer, I spend a lot of time in code review. Either I am reviewing someone else's code, or someone else is reviewing my code. In addition to making sure syntax is correct, that the tests are testing for the things they are meant to, and other items, one of the most important things to check for is overly-complicated code. Overly-complicated code can be too long, and it can be too short. It is not about the number of lines used to write a method, but the simplicity of the method itself.

Good code does not obfuscate its intentions. Good code can be read and understood without too much mental anguish and pain. Good code employs complex procedures only when absolutely necessary. Sometimes, it is easier to write more complex solutions to complex problems than to spend the time and compose a simple answer to a complex problem.

During the week of Passover, one of the central commandments of the holiday is to remove all traces of leavened products from our lives. We clean our cupboards, scrub our countertops and ovens, vacuum our cars, and more. But, do we also spend the time to remove the leavening from our hearts and our minds as well?

There is a note I keep on my desk at home with only one word in Hebrew: "פשוט - Simple." It is a reminder to myself to try and keep things simple and to not let matters become overly "leavened." The matzah consumed on Passover is among the simplest food items we will put into our bodies all year.

As we ingest our matzah this week, may we also internalize the call to simplicity, to keep things from leavening too much, and getting out of hand. Keeping things simple in our code. Keeping things simple in our emails. Keeping things simple can be liberating and a real boost of positivity in our interactions and in our lives.

2

SHAVUOT
A DIFFERENT PERSPECTIVE

This coming week is Shavuot, the holiday when we commemorate the revelation on Mount Sinai and the giving of the Torah.

The Torah describes this monumental occasion in vivid detail. But upon closer inspection, one verse seems a little strange (Shemot 20:15):

וכל־העם ראים את־הקולת ואת־הלפידם ואת קול השפר ואת־ההר עשן

"And all the people saw the voices and the torches, the sound of the shofar, and the smoking mountain…"

What does it mean for people to "see the voices… and the sound of the Shofar"?

Rabbi Akiva, one of the greatest Talmudic sages, explained that a miracle occurred and the Jewish people "saw that which is [usually] heard, and heard that which is [usually] seen," but that explanation raises more questions than it answers. What is the point of such a miracle? Why was seeing that which is heard and hearing that which is seen a necessary part of the giving of the Torah?

In data science, we place great importance on how we present and visualize our data. The way we present our data can have a significant impact on how we understand and consume it.

What can seem like a mess of disjointed coordinates becomes an elegant curve when plotted on a graph.

What can look like a bunch of random numbers (say, 6 54f 20 68 61 72 68 63 65 54 20 26 26 for example) when encoded in a slightly different way can convey the name of a great learning resource.

Sometimes, just changing the way you look at the same data can change its meaning drastically.

There's a story about the famous Chassidic master, Rabbi Levi Yitzchak of Berditchev (1809–1740) known for his nickname "the advocate of the Jewish people" for his hobby of spinning even the most seemingly negative traits in a positive view once told his students:

> "I don't know what God wants from the Jewish people; had he put the world to come in front of their eyes and the current materialistic world in books, there is no doubt that everyone would only pursue lofty spiritual goals. Instead, God put materialism in front of their eyes, and spirituality and the world to come he placed in books, is there any wonder that Jews struggle?"

Chassidic philosophy explains the difference between that which is "seen" and that which is "heard" as follows: something you see is something you experience directly, something you hear is something you can only experience indirectly.

For most of us living our lives in the physical world, physicality and materialism is the stuff we "see" and experience directly. Spirituality and

Godliness is something we "hear" and learn about, but don't experience directly in our day-to-day lives.

During the revelation at Mount Sinai, the Jewish people experienced God at such a direct level, that spirituality and Godliness became things that they "saw" and experienced in a very real sense. Meanwhile, the physical materialism of this world, while still existing, was so far removed from their direct experience that it was relegated to the realm of things that are "heard."

May we all merit to experience just a brief moment of that clarity this year, and may we accept and internalize the Torah this Shavuot with joy!

3

SUKKOT
EMBRACE FRAGILITY

How often are we greeted by advertisements telling us we must be strong? How often are we told that our goal as developers, DevOps engineers, or IT professionals is to make sure nothing ever fails? It seems we are inundated with the ideal of absolute strength as the prime virtue everywhere we turn.

However, Sukkot presents a very different picture. The holiday we are in the midst of during this Shabbat forces us to confront our fragility. In Israel, we joke that every apartment comes built with an armored safe room and also an open-air balcony on which to build flimsy huts. This is because, on the holiday of Sukkot, we transition our lives out of the stable, secure, and seemingly immovable walls we have built for ourselves, and move into the flimsy hastily constructed sukkah, or hut, that we have built for ourselves.

For one week, we are reminded that no matter how many redundancies we have built into our systems, no matter how many AWS clusters we have, or how many integration tests we have in our testing suite, things are always fragile. Things are fragile because life itself is fragile. We constantly make it appear stable, secure, and impenetrable, but that is just the facade we put on top of it to give ourselves some semblance of comfort.

Rather than this realization causing despair, it can be freeing. Once we reckon with the fragility all around us, we can embrace the fullness of our lives and our work. The fullness of our life and our work means that it includes moments of failure and of unexpected turns along the way. Knowing that these will happen can help us be better prepared for when they do happen.

Sukkot comes on the heels of Yom Kippur, when we aspire to never err again, because it is there to remind us that we will actually err again, and that's okay. We can aspire for absolute strength and perfection knowing that it is unattainable, but in the pursuit, we can reach our best selves, even if it is not our perfect selves.

4

EPILOGUE:
ONE YEAR AND BEYOND...

(One year of Torah && Tech concluded on the Parsha of Miketz and also on the week of the passing of Ben's mother, which was two weeks after the passing of his father.)

Two weeks ago, we were saddened by the news of the passing of Ben's father, Dan Pesach ben Baruch Shmuel Hakohen Greenberg. This week, after nearly 40 years of marriage, it breaks our heart to notify you that Ben's mother, Elisheva bat Chaim, joined her husband in eternal rest.

Our hearts go out to the Greenberg family, may they be comforted soon together with the mourners of Jerusalem, and may we only share good news from now on.

We dedicate this week's newsletter in her memory.

Chanukah is one of the most well-known, and well-loved, holidays in the Jewish calendar.

The name Chanukah has two meanings.

The first comes from the root חנך and means "renewal" and "dedication." That name comes from the rededication of the temple (the Chanukat Habayit) after the Syrian Greeks defiled it.

The second meaning of the name Chanukah is a portmanteau of the Hebrew words חנו כ"ה (they rested on the 25th) referring to the calm that was finally achieved with the end of the battles between the Maccabees and the Hellenists that occurred on the 25th of Kislev.

About a year ago, two rabbis turned web-developers, were trying to come up with ideas on how to keep in touch after an upcoming move and career switches that threatened to separate them.

A few weeks later, on the week of Chanukah, we realized the first meaning of the holiday's name by "dedicating" a new weekly newsletter, and that's how Torah && Tech was born!

Now, a year and 55 issues later, we are truly amazed at the impact we've been able to have. What started as a way of forcing two friends to keep in touch ended up touching 150 people a week! The feedback we've received from Jews and non-Jews, techies, and the "technically challenged" alike let us know that this is content the world really needs.

As we're getting ready to transition to the second meaning of Chanukah, as Torah && Tech is maturing and "settling down," the tech world still has many questions it needs to answer and conversations it needs to hold. Questions around ethics, morality, and the impact we have on the world.

At Torah && Tech, we are here for these conversations. We want to hear your voices. And we will continue to spread the voice of the Torah and be a "light unto the nations."

5

ABOUT THE AUTHORS

Rabbi Ben Greenberg served in different positions of Jewish communal life for over a decade before transitioning to a career in tech. He was the Orthodox Union and Hillel campus rabbi and Orthodox chaplain for Harvard University, a synagogue rabbi in Denver, Colorado, and worked in strategic planning at the UJA-Federation of New York. He has written extensively for publications such as *MyJewishLearning*, *The Boston Globe*, *The Denver Post*, *Time Magazine*, *The Huffington Post*, and others. He is also the editor of *Twitter Torah* and author of *Covenantal Promise and Destiny*. He currently works as a developer advocate at Vonage and lives with his family in Israel. He can be reached at www.bengreenberg.dev.

Rabbi Yechiel Kalmenson received his rabbinic ordination at the Central Lubavitch Yeshivah in New York. He traveled the world volunteering at various Chabad Houses and yeshivahs and served for several years as an assistant rabbi in North Woodmere, NY while teaching Judaic Studies at the Jewish Online School. He currently works as an engineer at VMware and lives with his family in New Jersey. He can be reached at www.yechiel.me.

CPSIA information can be obtained
at www.ICGtesting.com
Printed in the USA
FSHW022256080620
70733FS

9 781735 021706